e-Patent Strategies

for
Software, e-Commerce,
the Internet, Telecom Services,
Financial Services, and
Business Methods
(with Case Studies and
Forecasts)

Companion Volume
to
*Patent Strategies for Business
Third Editio*

Stephen C. Glazier
Washington, D.C.

LBI Institute
Washington, D.C.

... Stephen C. Glazier and LBI Institute, Inc.

Published by

LBI Institute, Inc.
P.O. Box 753
Waldorf, Maryland 20604

LBI Institute, Inc.
Suite 1800
3202 Rowland Place NW
Washington D.C. 20008

No part of this publication may be used or reproduced in any manner whatsoever by any electronic or mechanical means (including photocopying, recording, or information storage and retrieval) without permission in writing from Stephen C. Glazier and LBI Institute, Inc., except in the case of brief quotation embodied in news articles, critical articles, or reviews. The views expressed in the publications of LBI Institute, Inc. are those of the author and do not necessarily reflect the views of any law firm associated with the author, or of the staff, officers, or trustees of LBI Institute, Inc. This book is for general informational purposes only, and is not a substitute for professional legal advice. This book does not offer a legal opinion about any specific facts. Competent legal advice should be sought before any action is taken in a specific situation by a particular party.

Earlier versions of some parts of this book may have appeared as articles in various magazines. They are reprinted here with permission or reservation of rights. We would like to thank these magazines for their assistance. They include: *Artificial Intelligence in Finance*, *Managing Intellectual Property* (a Euromoney plc magazine), *InfoText*, *The Expert Witness*, *The International Business Lawyer*, *German-American Trade*, *The Legal Times*, and the *Journal of the Japanese Institute of International Business Law* (in Japanese translation).

Cataloging in Publication Data

Glazier, Stephen C.
    e-Patent Strategies for Software, e-Commerce, the Internet, Telecom Services, Financial Services, and Business Methods (with Case Studies and Forecasts)/Stephen C. Glazier.
    Includes index.
    ISBN 0-9661437-7-9
    1. Patents. 2. e-Commerce. 3. Software-Protection. 4. the Internet.
I. Title.

Library of Congress Catalog Card Number: 99-99126
ISBN: 0-9661437-7-9

Printed in the United States of America. Softbound.

Books by Stephen C. Glazier

Patent Strategies for Business
Third Edition

e-Patent Strategies
for
Software, e-Commerce, the Internet,
Telecom Services, Financial Services,
and Business Methods
(with Case Studies and Forecasts)

e-Patent Strategies
for
Software, e-Commerce,
the Internet, Telecom Services,
Financial Services, and
Business Methods
(with Case Studies and
Forecasts)

Companion Volume
to
*Patent Strategies for Business
Third Edition*

Contents

About the Author ix

Preface to this Volume:
Profits from e-Patents, i-Patents,
Service Patents, and Business Method Patents xi

What the Reviewers Are Saying
About The Companion Volume xv

To Order The Companion Volume xvii

To Order This Book xix

Preface to the Third Edition xxi

Preface to the Second Edition:
Patents are Business Tools xxiii

Acknowledgments xxix

A. Case Studies

1. Patent Due Diligence for
   Financing Technology Companies:
   Five Case Studies 1

B. *Software and the Internet*

2. e-Patents for Software, e-Commerce, the Internet, Telecom Services, and Financial Services: the *State Street Bank* Case and Business Methods 23

3. e-Patents and Business Methods, the Beat Goes On: the *AT&T v. Excel* Case 39

4. e-Patents for Internet and Banking Services: The Survey 45

5. Rules of Virtual Genius: Software and Internet Update 51

6. Four Stages of Patent Denial In the Software Industry 61

C. *Patent Audits*

7. Intellectual Property Audits: Start with the Business Plan 67

8. Intellectual Property Audits: Special Steps 73

D. *Technology Forecasts*

9. Tech Trends: Telecom Services 79

10. Tech Trends: the Internet, Intranets, and Extranets 83

e-Patent Strategies
vi

11.	Tech Trends: E-Commerce	89
12.	Tech Trends: Smart Cards	97
13.	Tech Trends: New Devices and Software	99

Appendices:

1.	*State Street Bank v. Signature Financial*	103
2.	*AT&T v. Excel*	125
3.	e-Patents: Seven Examples for Corporate Financing	145
4.	e-Patent Invention Disclosure Form	177

Table of Authorities 187

Index 189

About the Author

Stephen C. Glazier practices law in Washington, D.C., regarding patents, copyrights, trademarks, venture capital, licensing, and related business transactions for technology companies. He has a B.S. and an M.S. from the Massachusetts Institute of Technology (M.I.T.), and a J.D. from the University of Texas. He is a member of the bar in New York, California, Texas, and the District of Columbia, and he is a registered patent attorney with the U.S. Patent and Trademark Office.

Mr. Glazier holds six U.S. patents, in which he invents around prior art patents (see Chapter 5 herein, and Chapters 2 and 3 of *Patent Strategies for Business, third edition*, the companion volume to this book).

Mr. Glazier lectures frequently on legal and business topics, and is a contributor on legal subjects to the editorial page of *The Wall Street Journal* and other publications.

Mr. Glazier is a partner in the law firm of Pillsbury, Madison & Sutro LLP, and is a member of the Cushman, Darby & Cushman Intellectual Property Group of that firm. He may be contacted in Washington, D.C. at e-mail: glazier@alum.MIT.edu or glazier_sc@pillsburylaw.com, telephone: 202-861-3056, fax: 202-822-0944.

Preface to this Volume

Profits from e-Patents, i-Patents, Service Patents, and Business Method Patents

  e-Patents are patents for software and e-commerce. e-Patents include patents for services enabled by software. e-Patents for services include new telecom services, new financial services, and new ways of delivery of old services.

  i-Patents are e-Patents for Internet applications, including e-commerce.

  We invented these terms to describe the business tools, e-Patents and i-Patents, that this book teaches to exploit to pump up the bottom line of businesses.

  All these sorts of patents are often now lumped together as "business method" patents (although this "business method" terminology arises out of quirks of legal history, and may be a bit misleading.)

  Most businesses now have opportunities to use software and the Internet to increase their profits. In this sense, most companies are,

Preface to this Volume

or can be, software and Internet companies. However, most such companies will never be thought of as software or Internet companies, and will not make their money selling software or Internet applications. Instead, they will use software on the Internet to sell something else, probably goods, services, or information (or maybe even software).

When software on the Internet does something new, or does something in a new way, it may be patentable. This opens whole new strategies for patent profits, for a variety of industries that never could exploit the Patent Office before.

And when the e-Patent or i-Patent enables a new service, then the patent is, in effect, a service patent. (Should we call them s-Patents?) This is causing a stunning re-writing of business plans for large established businesses, and start-ups, in financial services, telecom services, insurance, and retailing of goods. You can now patent, in effect, financial products, subscription telephony services, insurance contracts, virtual stores, and many other business plans.

This book is an all new companion volume to the popular *Patent Strategies for Business, third edition.* This new book has chapters about e-Patents and i-Patents, including new rules to invent software and Internet applications (and to patent the result), current legal developments, intellectual property audits, related forecasts of trends in technology and business, and case studies of the use of patent strategies to make money.

There is tremendous money to be made by connecting the new chapters on technology forecasts, with the new (and old) chapters on how to invent and patent.

Current plans are to expand the section on case studies of technology transactions to become a separate volume of case studies. (They are lucrative, and fun.)

>Stephen C. Glazier
>Washington, D.C.
>27 November 1999

WHAT THE REVIEWERS ARE SAYING ABOUT THE COMPANION VOLUME

Patent Strategies
for
Business
Third Edition

"Stephen Glazier's book entitled *Patent Strategies for Business, Third Edition*, is a field manual for the intellectual property strategist to start thinking and acting... Glazier's book is one of the few sources which makes the effort to approach the patent field as a matter of strategy rather than as a matter of... how the authorities line up on each particular legal issue... Glazier's book lets the reader understand in a brief and manageable way how things work in the patent field... The writer wishes Glazier would convert his book into a multi-volume loose leaf series for which there is surely strong need and probably no better potential author or editor."

-Intellectual Property Rights News
Volume 2, Number 3
(Winter 1998)

TO ORDER THE COMPANION VOLUME

Patent Strategies
for
Business
Third Edition

Patent Strategies for Business, third edition, by Stephen C. Glazier, is the 454 page hardcover companion volume to e-Patent Strategies. To order a copy of *Patent Strategies for Business, third edition*, photocopy this page, fill in the information requested (please print), and transmit to the address indicated with your check for payment in full. The price of a single copy is US$99 (includes U.S. shipping and handling).

Name_____
Title_____
Company_____
Address_____
_____
_____

Mail to:
LBI Institute, Inc.   or   LBI Institute, Inc.
P.O. Box 753   Suite 1800
Waldorf, Maryland 20604   3202 Rowland Place NW
                                      Washington D.C. 20008
Or see on the Web:
    http://www.amazon.com/exec/obidos/ISBN=0966143795

for faster service, call: 301-645-0188 (USA)
or e-mail: LBI@TASCO1.COM   or fax: 301-843-0159 (USA)

TO ORDER THIS BOOK

e-Patent Strategies
for
*Software, e-Commerce,
the Internet, Telecom Services,
Financial Services, and
Business Methods
(with Case Studies and Forecasts)*

To order a hardbound copy of this book, photocopy this page, fill in the information requested (please print), and transmit to the address indicated with your check for payment in full. The price of a single copy is US$49 (includes U.S. shipping and handling).

Name_____
Title_____
Company_____
Address_____
        _____
        _____

Mail to:
LBI Institute, Inc.   or   LBI Institute, Inc.
P.O. Box 753   Suite 1800
Waldorf, Maryland 20604   3202 Rowland Place NW
                                      Washington D.C. 20008

Or see on the Web:
    http://www.amazon.com/exec/obidos/ISBN=0966143787

for faster service, call: 301-645-0188 (USA)
or e-mail: LBI@TASCO1.COM   or fax 301-843-0159 (USA)

xix

Preface, Third Edition
Patent Strategies for Business

Patents Are Business Tools

"Patents add the fuel of interest to the spark of genius."
-Abraham Lincoln

Popular demand for the second edition of this book encouraged us to reprint the second edition in an improved format with expanded distribution and a lower price. However, in working on the reprinting, we could not resist making additions and revisions. The changes quickly got out of hand, and as a result, the volume in your hands is actually now the third edition.

The first twenty-five chapters are largely unchanged, although there are many small improvements, updates, additions, and corrections. Particularly interesting changes were made in Chapter 3 ("Invent Around Your Competitor's Patent, and the Antidote"). Chapter 13 ("New Developments 1995") should now be read in light of Chapter 31 ("New Developments: 1996 and Early 1997"), particularly regarding the Patent Office software patent guidelines. And because of their importance, the PTO's new software patent guidelines are added in their entirety as Appendix 3.

All of Section VIII, beginning at Chapter 26, is totally new.

Preface, Third Edition

Again, let us emphasize that this book is not a "dumbed-down" or "beginner's introduction" to patent law. Instead, it is a sophisticated discussion of selected key approaches to make intellectual property serve a company's business plan and goals. The targeted reader of this book is a CEO or general counsel of a U.S. or foreign corporation, whether large or small. Patent lawyers have also found this book useful to clarify a client oriented view of the patent profession.

A word of caution: Nothing is quite as simple as it first sounds. Of necessity, many complicated legal technicalities are glossed over or not discussed in this book, in an effort to communicate fundamental strategies. These strategies are valuable in many cases, but may be unavailable in other cases. Any such unavailability may arise from special facts, surprises, financial constraints, changes in the law, or otherwise. (Also, any part of this book may be obsolete or otherwise in error at the time you read it.) And, of course, it seems that any useful legal statement, and certainly any evaluation or judgment of merit, is arguable. So please consult legal counsel, your special facts, and current law, before assuming that anything herein may work or be advisable in any particular case. (But let's not lose the forest because a few trees change colors; it is our expectation that the big ideas in this book will survive many changes in the details of law and business.)

The reader is encouraged to communicate to the author any comments regarding preferred additional topics for the fourth edition of this book. Thank you for your interest.

> Stephen C. Glazier
> Washington, D.C.
> 2 September 1997

Preface, Second Edition
Patent Strategies for Business

Patents Are Business Tools

> *"Congress shall have power... To promote the Progress of Science and useful Arts, by securing for limited Times to Authors and Inventors the exclusive Right to their respective Writings and Discoveries."*
> U.S. Constitution, Article I, Section 8

This book is written as a practical guide to the use of patents as effective business tools. That is, this book is written for business people and attorneys who are not intellectual property specialists, but who do have opportunities that can be pursued by practical patent strategies. Other areas of intellectual property law are also touched upon, particularly regarding copyright and trade secrets, where they apply to software.

This book is designed to be a tool for the reader, to present ideas and strategies that have been successful for others, and that will be successful again. This book is not an academic treatise or survey of the entirety of patent law for the patent attorney specialist. It should, however, be useful for the patent attorney, since the book explicitly discusses important strategies and concepts that are seldom analyzed in print. Current legal developments and their practical applications are also discussed.

The approaches in this book have grown out of specific transactions with specific parties, so we are confident that the ideas in this book have passed at least their trial runs in the real world. However, we have been careful to keep all the discussions here on a generic level, so that no confidences regarding any particular party are revealed. Any references to specific companies or individuals in this book communicate only information that is disclosed in the public record, and which involves parties that this author has never personally represented.

There are several major trends in patents in the last ten years that have revolutionized the patent business, and which we expect to continue for the foreseeable future.

Patents were made more enforceable, and appellate law uniform and more correct by the creation of the Federal Circuit Court of Appeals in the mid-'80s. This arose out of a general recognition that the U.S. was in a competitive world marketplace that it could no longer dominate, and hence the U.S. had to protect its intellectual property in order to remain competitive. This economic situation will not change, so we can count on this legislative change in patent law to be permanent.

Also, software and computer applications have been determined to be patentable technology, if they are new and not obvious. This legal development takes place as software becomes more prevalent in daily life, and as custom circuits are being replaced as a design approach in favor of standard chips with custom programming. At the same time, hardware becomes cheaper, for processing and for memory. The result is that in more and more industries, more of the innovation is in the form of patentable software applications. This has lead to the shocking application of patents to new areas such as telecommunication <u>services</u> and financial <u>investments</u>. Also, we may be about to see the rise of business method patents, to a par with the current patentable status of all other man-made processes and methods. At the least, we can expect for software patents to become a factor in most industries in this decade.

Patents are not just for gizmos anymore.

We are also seeing the globalization and harmonization of patent law across our planet. This has lead to the currently pending amendments to the U.S. intellectual property statutes that arose from the GATT treaty, and these amendments look minor compared to what we may see in the next five years. This trend has also lead us to spend more time in Munich (at the European Patent Office) and Tokyo (at the Japanese Patent Office) than we ever would have guessed, and to the fact that about half of our clients are foreign to the U.S.

These developments have greatly increased the power and value of patents. Consequently, patent questions are now critical in many commercial and financial transactions. Judgments for patent infringement are bigger (Kodak recently paid Polaroid almost $1 billion for patent infringement, and Litton more recently received a judgment against Honeywell for $1.2 billion), and patents are more often enforced when litigated. It has also been reported that Texas Instruments has received more revenues in some recent years from its patent royalties than from the sale of its products. This has put more pressure on management to develop their patent portfolios, and to avoid infringement. It has also put pressure on Congress to fix the ownership section of U.S. patent statutes (see the appendices herein for one proposed bill).

As a result, financiers are now beginning to do real patent due diligence, and are looking for patent attorneys who can do a deal.

Preface, Second Edition

Also, with the possibility of triple damages for patent infringement, we have seen the beginnings of a plaintiff's contingency bar for patent infringement, and an increasing ability of small patent owners to negotiate licenses with major corporations. Also, we have seen the beginning of an evolution away from the patent specialty law firm, to the patent group at a general practice law firm. We have also seen the first younger transactional patent attorneys.

With the first appearance of biotechnology companies, the courts have agreed that man-made life forms are patentable. (This development is not dealt with in this edition of this book, and is a matter for the third edition.)

We have also seen that copyright can be used to obtain some protection for software. However, copyright protection is weak compared to patent protection for software (when you can get a patent). The unique overlapping nature for copyrights and patent rights for software require that both sides of this ownership coin be attended to for software.

We have also seen case law develop and clarify the legal side of inventing around a competitor's patent (a process of legal development which is still continuing). This makes the invent around process more reliable and practical. Also, this in part enables what we personally find to be the most exciting innovation of this book, which we call the Rules of Virtual Genius.

Inventing and patenting are two different things that are part of a larger commercial effort, to make money from new products and services. Surprisingly, current developments in technology and law are actually working in the same direction, so that good inventing can facilitate good patenting, and, most unexpectedly, good patenting can facilitate good inventing. We will probably never be able to advise, in general, to go to a patent lawyer <u>for</u> an invention (instead of <u>with</u> an invention); however, inventing and patenting are a lot easier than most technical people realize, especially if they stay focused on the Rules of Virtual Genius.

We also have found that the Rules of Virtual Genius are teachable to most business and technical people, and lessons on this subject can inspire profitable innovation.

We also have seen since the late '80s that applications at the U.S. Patent Office have been growing at about 15% per year, while software applications have been increasing at about 30% per year.

In the five years that I was at M.I.T., we spent a lot of time developing and applying new ideas, but I cannot recall ever hearing the word "patent" mentioned. I certainly do not recall a course on the subject. M.I.T. is an excellent place and probably the best institution of any kind that I have been involved with, and I hope that the approaches in this book can be of some use to the people there, and others similarly situated.

In effect, this is the book (or at least part of it) that I wish someone else had written for me when I was an engineering student in the '60s at M.I.T., or a law student in the '70s. Hopefully, it can now give others a faster start and a better direction in the '90s.

The point of this book is to help businesses make money. It does this by pointing out patent strategies and new legal developments that may offer profitable opportunities, in specific circumstances, within a company's budget and schedule.

Stephen C. Glazier
Washington, D.C.
23 October 1995

Acknowledgments

My clients over the years have pressed me to invent and organize the ideas found in this book. Ultimately, those clients gave these ideas their greatest seal of approval: they used them. I would like to give my clients, past and present, special thanks.

(However, I would like to tell Liz, Karen, Javier, Joe, John, Tom, Barry, Paul, Gerry, Ted, Jim, Mike, Gene, Tim, Heinrich, Anders, Dexter, Jud, and Chuji that they can relax. My best ideas remain in my unpublished confidential correspondence to them.)

Hi, Hawkwood! Hi, Jody! Hi, Mom!

1

Patent Due Diligence For Financing Technology Companies: Five Case Studies

"I do not take business risks; I control them."
　　　　　　　　　　　　　　　　-J. P. Morgan

"Chance favors the well-informed."
　　　　　　　　　　　　　　　　-Marcel Proust

"My best ideas are somebody else's."
　　　　　　　　　　　　　　　　-Benjamin Franklin

　　　Due diligence for intellectual property is becoming increasingly important in mergers and acquisitions, venture capital, IPO's (initial public offerings), and other financial transactions for technology companies. For technology company deals, patent due diligence can make or break a deal, or radically change the price.

This is due to several factors. One factor is the increased market capitalization of technology companies that have always viewed intellectual property, particularly patents, as critical to their market share and profit margins. These companies, including now software companies and Internet companies, are actively engaged in mergers and acquisitions. Also, the telecom world outside of the Internet is being changed by a combination of de-regulation, mergers and acquisitions, and a very dynamic technology infrastructure (including wireless, the Internet, convergence of cable and telephone, PCS, satellite-base wireless, and electronic commerce). Similar growth and M&A activity in the entertainment industry, where copyright of content and trademarks, are the crown jewels, have also brought intellectual property due diligence to M&A.

Another factor contributing to the importance of intellectual property due diligence in technology finance, is the increasing number of industries, some long established, that are newly coming to the patent world. In the United States, it has now been well established (in this decade) that software developments are patentable. Also, there is a new federal appeals case, *State Street Bank & Trust Co. v. Signature Financial Group, Inc.*, 149 F.3rd 1368, 47 U.S.P.Q. 2d 1596 (Fed. Cir. 1998), that clarified that even software enabled financial services are patentable. Hence, now we have at least two new industries, financial services (including banks, securities brokerages, insurance companies, savings and loans, asset managers, program traders, mutual fund managers, and others) and telecom service providers (as opposed to telecom equipment manufacturers) that are moving actively into the patent arena for their software-enabled services. These are two large established industries with huge market capitalizations that are consolidating and reorganizing with extensive M&A activities, which have newly found themselves dealing with patent crown jewels for new market sectors.

Case studies from practice can be looked at to shed light on patent due diligence in corporate transactions. Lessons can be drawn from these practice examples about the best ways to pursue these activities. (As they say in the old black and white detective movies, the names have been changed to protect the innocent in some of these case studies. Where specific names of parties are mentioned, all the information given has been drawn from the public record, and no client relationship has existed between this author and the indicated parties.)

Case No. 1: *Due Diligence as Industrial Espionage, or "My Best Ideas are Somebody Else's"*

Regardless of the original intent of the parties, any due diligence project can in retrospect appear to have worked like a sophisticated commercial espionage project against a competitor. This can be fine for the party that gains the information, if it violates no law and breached no contract. However, for the party that teaches in this process, it must be a no win game.

Take this scenario. Big Fish Co. expresses an interest in investing in, buying out, joint venturing with, or using as a supplier, the aspiring Mullet Co. Mullet Co. is of considerable interest because it has developed a wonderful new fish hook early warning system, but does not have the capital to market it outside the metropolitan area of Pikes Peak, Colorado. But before Big Fish can make a decision on Mullet, Big Fish must do its due diligence and become an expert on Mullet and its product.

In the due diligence process, Mullet shows Big Fish its books, and the software algorithm that makes the hook alarm so unique and effective. Big Fish loves the algorithm, which is so brilliant that it requires only four lines of source code. But Big Fish walks the deal because its bylaws require that all members of the Board of Directors have a driver's license, and this disqualifies the President of Mullet Co., until he has two more birthdays. And Mullet Co. insists: no Board seat, no deal with Big Fish.

Six months later, Big Fish is on the market across the continent with a hook alarm that works almost as well as Mullet's and costs twice as much. But Mullet Co. is chopped into fish meal by Big Fish's huge sales force.

Was all this a sophisticated spying job by Big Fish, or an unavoidable development? After all, most prospective deals never happen. In fact, the answer to the question means little to the parties, except as it may affect their legal rights.

(But there is a happy ending for Mullet Co. Mullet has one of those new software patents for its algorithm, and sued Big Fish for patent infringement. To Big Fish's surprise, Mullet won and received a nine figure judgment and a permanent injunction against infringement. Big Fish starts shipping its hook alarms without software and sales crash. Within a week, Big Fish is back at the negotiating table with Mullet. Mullet's founders sell out for a huge price, and they all go fishing.)

The tension in due diligence is that the cooperative target must treat due diligence as an actual espionage activity, but still try to get a friendly deal done. Non-disclosure agreements can be obtained, but they are only contracts that can be breached, and for which any defendant may have defenses. The Uniform Trade Secrets Act found in many states may give some recourse, if the aggrieved party survives. Also, the new federal *Economic Espionage Act of 1966* may give some relief. (See chapter 26 of *Patent Strategies for Business, third edition*, the companion volume to this book, for a detailed discussion of this topic.) Perhaps the best course for the target is to reveal as little as possible, and reveal no technical product information at all that is not in pending or issued patents.

In the above scenario, if Big Fish is in fact using the Mullet algorithm, Mullet will be best protected if the algorithm is claimed in valid patents owned by Mullet. As Stac showed in its litigation discussed below, only this strategy can make Mullet a piranha, instead of chopped bait.

Case No. 2: *Stac v. Microsoft, or "Are these Guys Protected?"*

"6/08/94 --   PERMANENT INJUNCTION [against future infringement] against defendant Microsoft Corp...
6/13/94 --   AMENDED JUDGMENT AND ORDER ... Microsoft shall pay Stac $120,000,000.00 ... "
- Civil Docket for Case # 93-CV-413
Stac Electronics v. Microsoft Corp.

"PLEASE READ THIS NOTE CAREFULLY. MS-DOS 6.21, the version included with your new PC, does not include the DoubleSpace compression utility. The documentation you received includes references to the DoubleSpace feature. Please disregard any reference to DoubleSpace in the accompanying documentation. We apologize for any inconvenience this may cause you."
- *From a sticker on the back of the MS-DOS 6.2 Manual, shipped <u>after</u> 6/8/94. The "DOS Lobotomy by Stac" Notice.*

The preceding parable in Case No. 1 may show some relation to the fact pattern in *Stac Electronics v. Microsoft Corp.* (D.C. C. Cal. CV-93-413-ER, 1994). In this suit a small firm with a software patent (Stac Electronics, which had a patent on the algorithm for its PC hard disc data compression software product) defended its market niche in court against a large infringing corporation (Microsoft).

The lesson for the patent buyer, inventor, or underwriter, from *Stac Electronics* and Case No. 1 above, is that the target must be carefully reviewed to ascertain if its products are adequately protected by intellectual property strategies. Only then can the target be protected from copying or reverse engineering.

The second lesson, from Big Fish's and Microsoft's point of view, is that a new product or service must be carefully reviewed to assure that it does not infringe a competitor's patent position.

e-Patent Strategies

Apparently, Microsoft had expressed an interest in working with Stac, and did due diligence on the Stac product. In this process, Microsoft apparently decided to copy the compression algorithm of the Stac product. Microsoft then wrote its own code to execute the Stac algorithm and used the code in the Microsoft DOS 6.2 product.

Stac sued Microsoft for patent infringement, copyright infringement, and trade secret violations. Stac lost on all counts <u>except</u> the patent infringement count, on which Stac won. Stac received a judgment of about $120 million, plus a permanent injunction against Microsoft to stop further infringement.

After the litigation, for about a week, a lobotomized version of DOS was shipped with the compression feature disabled. DOS manuals were shipped with stickers on the cover warning to ignore the chapter on compression, because the feature had been disabled (pursuant to the federal injunction). After about a week of this, Microsoft cut a deal with Stac by which Microsoft, apparently, paid Stac a large amount of cash, Microsoft made a large capital investment in Stac, and Microsoft received a license to use the algorithm in DOS.

Apparently, after getting caught in court by the patent, Microsoft thought it wanted to deal after all. And Stac found a price that looked good.

If not for its patent and the resulting injunction against Microsoft, Stac would likely be in serious financial trouble today, or out of business. How could Stac sell its product if the same features were available in DOS (or Windows 95) for free? Without a patent, Stac would have just been a free test market project for Microsoft.

Notice that Stac's corporate life was saved by its software patents only (which covered its basic algorithm). Software copyright got nothing for Stac because Microsoft did not copy source code;

instead, Microsoft apparently just wrote new code for the same algorithm.

This is the new paradigm. This is how a little software company can save itself from annihilation by a giant with overwhelming market power and financial strength.

This approach also applies to any easily copied product or service that can be protected by software patents. Financial products and services are a prime example of this. New financial products are now receiving patent protection by a species of software patent. This is important because a new financial product can be very quickly copied once it achieves market success.

Immediately after Stac's courtroom victory against Microsoft, the president of Stac announced to the public, correctly we think, that this was the new paradigm for the development and marketing of new software developments. That is, patents must be obtained for new software products, so that they can be protected from easy copying and infringement.

Note that this patent protection for software is superior to the traditional copyright protection. This is because copyright protects little more than copying of source code, and perhaps screen displays and user interfaces, while patents can protect the basic concept of a software product, regardless of the actual source code. In the case of *Stac Electronics v. Microsoft Corp.*, Microsoft avoided a copyright problem but ran afoul of patents.

The further development of this new paradigm can be seen in the very recent initiation of further internet software patent cases, including *Priceline.com v. Microsoft Corp., Amazon.com v. Barnesandnoble.com, Trilogy v. Carsdirect.com,* and the Yahoo.com litigation. In each of these cases, freshly issued software e-commerce patents went straight to enforcement litigation. It is expected that many more such patents are in the application pipeline at the Patent Office, and will lead to further developments to be watched with great interest.

Case No. 3: *Own It Before You Sell It.*

Godzilla Corp. was a big corporation with about 900 engineers on its staff, business activities in several high tech areas, and no patent portfolio or technology spin off operations.

At some point, Godzilla's brain, its upper management that is, got the brilliant idea that 900 engineers employed in high tech fields may occasionally have a new idea of value. Godzilla smelled money. Furthermore, with over $20 billion in revenues, the Big Lizard (as its employees called it, affectionately we think) felt that it could scrape up a few bucks to chase a good business opportunity.

Leaping into the field in a big way, Godzilla selected one new gizmo, the Big Can, got the Lizard lawyers to apply for and receive a patent on it, and began marketing the opportunity.

The pitch was good: Godzilla needed the Big Can, but nobody made the Big Can, so Godzilla had to design and manufacture the Big Can for itself. Now all the other little lizards in the business also needed the Big Can, but Godzilla was not a manufacturer, at heart, but a consumer of Big Cans. Of course, nobody else could make the Big Can, even for their own use, because 'Zilly had the patent monopoly. Unless, of course, you did a license deal with the Lizard. License the patent, hire the Lizard engineers as consultants to teach you how to make the Big Can, and then you would be the Big Cheese of the Big Can. It would be easy. The testimonials alone from the Happy Lizard would sell the product.

Godzilla spent $750,000 marketing the pitch to the major players. They had a license deal cut with Not So Big Can Corp., a can company with ambitions to become, well, bigger. Not So Big would give Godzilla seven figures up front (that is, several million dollars), and a percentage of the gross, for the life of the patent. Godzilla's

e-Patent Strategies

projections indicated about $100 million in royalties from the deal over the remaining life of the patent.

Godzilla was ready to party, the deal looked so good. The Liz was even considering filing another patent application (they still had 899 engineers left with untapped ideas).

After extensive negotiations, a huge and complicated license and joint venture agreement was negotiated between Lizard and Can. They were ready to sign, pass the check, and open the champagne.

Three days before the closing, Not So Big did some due diligence on Godzilla and the contract.

Stop the music.

The first thing Not So Big's intellectual property attorney did was to check title. That is, he checked the records to see who owned the $100 million patent in question. Whoops, no assignment was on file for that patent in the public patent assignment records. The legal details are involved, but basically Godzilla needed a written assignment for the patent from the individual named inventor to Godzilla (even though the inventor was an employee of Godzilla), for the $100 million deal to close. Now, Not So Big is in the business of making deals, not breaking deals. And the point of finding title defects is not to kill a deal, but to correct the defect before it kills the deal maker.

They tell me that the meeting with Godzilla and the inventor was rather ugly. At first, the inventor held out for too much money, so much that it would have killed the deal. Finally, the inventor settled for a large cash "early retirement" package, and signed the assignment to Godzilla. The deal was saved, barely, at considerable extra expense.

Patent Due Diligence

I am sure that on the next deal, Godzilla will save a lot of money, and very likely save the deal, by checking title to its patent before it develops and markets the product. I have found that an employee or consultant will usually sign away rights to intellectual property, on the front end, in order to initiate or maintain employment. However, on the back-end, after they see a big market developing, their hand can get paralysis. Feeding their checking account appears, at that point, to be the only cure for this strange disease.

Case No. 4: *Buying the Defendant's Seat*

Strip Co. had a machine that manufactured a particular part with a plastic strip laminated on it that could record information. Strip Co. wanted to sell this product line to concentrate on other lines of business. Strip Co. contacted Large Co., which was interested in making machines that made parts with plastic memory strips. The parties negotiated a contract providing for the sale "the product line" for a $15 million one time cash payment. Strip Co. had a patent portfolio of six U.S. patents that covered its "product line", provisions for assignments of which were included in the contract. Three days before the closing, Large Co. started a bit of patent due diligence.

Initially the Large Co. patent attorney had difficulty in translating "a product line" into any species of legal property. That is, the contract appeared vague as to the description of a fundamental term, that is, the goods or property being sold. Since the machinery in question was manufactured by Strip Co. only to order, no inventory was maintained. There was one purchase order for which equipment was being manufactured, so that the rights in that executory contract could be assigned together with the work in progress for that purchase order. Other than that, no property to be transferred could be identified other than the portfolio of six patents.

That then, a patent portfolio, was largely what Large Co. was really buying.

Considering the three day deadline for closing, Large Co. immediately checked title for the six patents. Counsel found that written assignments had been filed at the Patent Office by the inventor of record to his employer, Strip Co. only in the case of four patents. That is, the two other patents in the portfolio, according to the public assignment records, still remained the property of the individual named inventors, and not of Strip Co. That is, Strip Co. was endeavoring to sell the property for which there was a title defect, to put the matter optimistically.

In the interest of making deals rather that breaking deals, Large Co. demanded that the contract be amended to include (1) a covenant to convey title to the patent portfolio, (2) representations and warranties of good and unencumbered title to the patent portfolio, and non-infringement by the Strip Co. products of the patents of competition, and (3) various other representations and warranties. Furthermore, Large Co. demanded that the entire payment of the purchase price be placed in escrow pending verification of the representations and warranties, and performance of the covenants. The closing in three days would in effect be a conditional closing with the funds escrowed pending verification of title, non-infringement and other points. After much noise by Strip Co., the escrow of the purchase price was established.

Further activity during the escrow determined that the inventors of one of the "loose" patents were American citizens and residents, and still in the employ of Strip Co. Their assignments were obtained quickly and without incident (but it is not generally known at what cost).

However, the inventor and owner of record of the sixth patent was a Japanese citizen and resident, who at the time of the sale had advanced to the somewhat distant relationship with Strip Co., of being an ex-employee. Indeed, the inventor had retired and was thought to have moved to the northern Japanese island of Hokkaido. No current information was immediately available as to where (or whether) he still lived. After several months Strip Co. did produce an assignment to Strip Co. in English of all rights in the Japanese inventor's patent, with a Japanese signature and notarization.

During this extended escrow period while the assignment from Japan was being obtained by Strip Co., further due diligence was undertaken.

One aspect of the due diligence project was to obtain and review the major prior art patents related to Strip Co.'s patents. General first impressions of validity were obtained from this analysis.

Also, Large Co.'s patent counsel stepped back and looked at the body of the prior art to make general characterizations of the portfolio as it may appear useful to the transaction in any manner. It was determined that the prior art patents by competitors probably suggested the range of competing products in the marketplace. The prior art patents were grouped according to their owners, and it was concluded that the most significant portfolios of prior art owned by competitors were in fact owned by the two largest competitors of Strip Co. Furthermore, it was determined that there was a substantial possibility that the products represented by Strip Co.'s patent portfolio may be dominated by the patent portfolios of prior art owned by Strip Co.'s largest competitor.

That is, on the face of the patents analyzed, it appeared that if Strip Co. in fact manufactured and sold the products claimed in Strip Co.'s own patent portfolio, then Strip Co. may run a substantial chance of infringing the earlier prior art patent portfolio of its major competitor. That is, a major motivation for Strip Co. selling its product line at this time may be to avoid a possible infringement suit by its major competitor.

Large Co. determined to proceed with the acquisition, but as a result of the due diligence and the identified risk of infringement, the purchase price was discounted from $15 million in cash to $7 million, with $3 million paid in cash at closing and $4 million paid over time, provided that no infringement claims arose.

Hence, the patent due diligence on this project succeeded in closing the transaction, but points identified regarding title defects, patent coverage and potential infringement led to a material modification in the price terms and the general business evaluation of the acquisition target.

The lesson here for the seller is to clean up your patent portfolio, to raise it to investment grade, before the deal. If you can.

For the buyer, the lesson here is that the earlier you do the due diligence, the better. And look at title first.

Case No. 5: *Patent What You Sell, Sell What You Patent*

Needles Co. had a line of medical devices that had been developed by its founder and CEO, Mr. Needles. Mr. Needles was approaching his late fifties and wanted to sell his company, and spend the remainder of his career developing products for manufacture and sale by others. Needles Co. was headquartered in Australia but had sales throughout the world, with its biggest single market in the United States. Needles Co. approached a large U.S. medical device company, Big Medco, which liked its products. However, the evaluation was that Mr. Needles had little to sell other than the customer list of Needles Co. since none of the products were protected by patents. Big Medco could simply copy the devices of Needles Co. and manufacture without royalty payment.

At this point Mr. Needles went to U.S. patent counsel to seek a remedy. U.S. patent counsel determined that the devices currently on sale by Needles Co. could not be patented in the U.S. because they had been on sale for more than one year. However, the strategy was developed to immediately patent all products currently under development by Needles Co., and to aggressively pursue second generation improvements to the current product line where such improvements could themselves be patentable. Hopefully, then the current product line in the public domain could be made obsolete and replaced by second generation improvements that were patented and proprietary to Needles Co.

In this way it was hoped that Needles Co. could evolve from a public domain manufacturer to a proprietary patented manufacturer. This would then require Big Medco to buy Needles Co. on the basis of a multiple of its earnings, rather than simply for the value of its customer list.

Over the next six years a portfolio of U.S. patents and foreign counterpart patents were obtained by Needles Co. for over a dozen of its main products. Most of the patents were invented by Mr. Needles himself. In some instances, patent counsel co-invented certain improvements for the benefit of Needles Co., in what might be called, "legally driven" product development. (That is, product development and patenting driven by legal analysis of patentability opportunities developed from the related patent prior art. Traditionally, there are two approaches to new product development, "market driven" and "technology driven". "Legally driven" product development is a third possible approach to product development and innovation that is new and relatively unknown, but can be highly productive).

Furthermore, the developing patent portfolio of Needles Co. became a growing prior art bar to the later patent applications of Needles Co. itself. In response to this expected and natural trend, the patent applications of Needles Co. began to incorporate more and more software novelty, rather than merely hardware novelty. This coincided with the general trend towards "smart" medical equipment that developed parallel with the legal developments in the United States allowing for the patentability of software. Although a "medical device company", the highest profit margins for Needles Co. now came from sales of its patented software, and of unpatentable disposable parts needed in each use of the software.

While developing this patent portfolio, Needles Co. added to its proprietary product line by going public in Australia and acquiring (for stock) a competitor located in Italy with a patented product line. At the end of last year, Mr. Needles reapproached Big Medco and sold Needles Co. in a transaction for stock and cash, and at a valuation that was a material premium over the then publicly trading price for Needles Co. In his last communication with U.S. patent counsel, Mr. Needles informed counsel that he was purchasing a new house, rumored to be one of the largest private residences in Australia.

A lesson here is "everyone is a software company now".

Another lesson here is "sell what you can patent and patent what you can sell".

For the buyer, the due diligence lesson is to carefully review the products <u>and the software</u> for patentability. For the seller, the lesson is the same, but do it well in advance to develop an investment grade intellectual property portfolio.

Three Practical Tips

There are three general practical tips that we can offer here to help the patent due diligence process when reviewing possible investments in technology companies, and for managing the investment process if a positive financing decision is made.

1.   A Competitive Advantage. A large competitive advantage can be had in the finance of technology companies by those who obtain access to patent professionals who can add value to a financial transaction. This is true for several reasons. Clearly, some large companies have sophisticated in-house professionals who fully understand their intellectual property and how to deal with it. But, most companies cannot and do not have such talent on staff. Likewise, many operations that are financing technology companies, including some well known venture capital concerns, cannot and do not have the expertise to evaluate a patent portfolio, and its impact on a company. Furthermore, many patent law practitioners are not aggressive in pursuing these issues, for a variety of reasons which may include lack of client interest, lack of experience, and lack of personal interest.

2.   An Epidemic. Basic title problems are everywhere in the patent world. But they can often be cured if acted on early enough. Ownership issues are the first thing to review in a patent portfolio. But like poorly poured concrete, the longer they sit, the harder it is to fix them.

3.   Budget: Sixteen Items. Almost no deal has a budget or schedule that will allow all possible patent due diligence to be done. The idea is do the basics, and then use judgment and some creativity to react to the basic results and to ascertain what else to do.

For example, validity opinions are rarely done in patent due diligence, because of the customarily large expense.

Perhaps the following is a good minimum menu for patent due diligence:

(1) Properly identify all patents in the key transaction contracts, and use words of conveyance for these patents.

(2) Verify recorded assignments from all the inventors.

(3) Review all Patent Office filings and UCC filings for clouds on title.

(4) Include all foreign patents of interest.

(5) Review all licenses, partnership agreements, employment agreements, and other relevant contracts. Watch out for ownership "leaks" especially for "after developed" inventions.

(6) Do you need a title opinion letter for the patent from your patent counsel?

(7) Ascertain if the target's products fall within the scope of their patents.

(8) How secure are the target's trade secrets? Review its "confidentiality program."

(9) Has the target patented its software or its services?

(10) Review all the target's past attorney opinion letters regarding patents.

(11) Review all past "cease and desist" letters regarding infringement, both sent and received. Query if laches

has arisen? Is infringement willful? Is the target about to be sued?

(12) Investigate the competition's patent portfolio, U.S. and foreign. Get your patent advisor to characterize the forest, <u>not just the trees</u>. Where is the industry going, and how does the target fit in? Can the target protect its new products and services with patents?

(13) Get <u>representations and warranties</u> from the target, and its patent counsel, regarding all of the above.

(14) Once you have done the above, step back and decide what else should be done in this particular case, within the current schedule and budget. For example, can the target "invent around" its competition's patents? Can the target "invent-on-demand," or "invent-just-in-time," to evolve its currently unpatented products into new patentable variations?

(15) Remember, if you cannot cure a problem any other way, you can always change the price in the deal, set up escrows and offsets securing representations and warranties, or ultimately (and this is the fire escape of last resort) you can just not close. Litigation, of course, is not a business transaction, and is best avoided.

(16) Verify ownership of any overlapping, related, necessary or convenient patents, copyrights, or trade secrets, embodied in any proprietary software or other products or services.

2

e-Patents for Software, e-Commerce, the Internet, Telecom Services, and Financial Services: the *State Street Bank* Case and Business Methods

A recent case, *State Street Bank & Trust Co. v. Signature Financial Group, Inc.*, 149 F.3rd 1368, 47 U.S.P.Q. 2d 1596 (Fed. Cir. 1998), has unleased a frenzied rush for e-patents for software of all types. This includes patents for pure software applications, e-commerce, the Internet and Internet enabled services, telecom services, virtual retailing over the Internet, and other "business method" applications.

State Street Bank continues the pro-software patent line of cases of the Federal Circuit, including *In re Alappat*, 33 F.3rd 1526, 31 U.S.P.Q. 2d 1545 (Fed. Cir. 1992), *Arrhythymia Research Technology, Inc. v. Corazonix Corporation*, 958 F.2d 1053, 22 U.S.P.Q. 2d 1033 (Fed. Cir. 1992), and *Paine, Webber, Jackson & Curtis, Inc. v. Merrill, Lynch, Pierce, Fenner & Smith, Inc.*, 564 F. Supp. 1358, 218 U.S.P.Q. 212 (D. Del. 1983), the "CMA case". This case does not break radical new legal ground and merely continues the unbroken march of cases in the pro-software patent direction. However, *State Street Bank* has had an earthquake-like effect in several industries, because of the unprecedented press coverage it has received. This has been particularly true in the area of patents for new financial services enabled by software. The unusual publicity has probably been caused by the fact that the defendant in this case, State Street Bank, is the first financial institution which directly challenged the use of software patents to effectively patent new financial services, and clearly lost. The entire financial industry is now on notice that new software enabled services are indeed patentable subject matter, provided that sufficient levels of novelty are obtained.

Now for the first time, whole service sectors that have never developed patent strategies, particularly financial services and telecom services, must now pursue effective patent strategies for their new services and software/Internet delivery platforms, or place themselves at a severe competitive disadvantage. This is perhaps the first time that major developed industries, such as banking, retailing, insurance, and telecom services (such as long distance telephone companies and local telephone service providers, as opposed to telephone equipment manufacturers) are newly required to go to the Patent Office. Usually, when new industries appear on the patent scene, the markets and the companies themselves are yet to develop any financial size. But in this case, major multibillion dollar industries, with huge developed markets, are now newly throwing themselves at the Patent Office with patent strategies that are well funded from inception.

Financial Service Providers

The race is now on for banks, insurance companies, securities brokerage firms, savings and loans, asset managers, program traders, mutual fund managers and other financial service providers to obtain a patent monopoly on their new products and services where they are implemented by software.

The floodgates have been opened on this new trend by a new Federal Circuit decision that aggressively breaks new ground to enforce patents for new financial products and services, and for other methods of doing business that are implemented by software. (Indeed, this case continues the larger legal trend to promote patents in all areas of software applications.)

The financial services industry had traditionally been completely uninvolved in the world of patents. But it will now be critical to the development of market share and cash flow in new product niches to protect a competitive advantage using patent strategies. It is rare for mature industries to enter the patent field late in their history. However, this is now happening in the financial services area with changes in the law that permit patents for software implemented innovations in this market. Banks, insurance companies, securities brokerage firms, savings and loans, asset managers, program traders, mutual fund managers and other financial service providers are now beginning to apply at the U.S. Patent Office and to actively license some of their new products. However, most of the financial players are still not active with patents and have not picked up on their new patent opportunities. Inaction in this field will, of course, lead to loss of competitive advantage and market share in new product areas. Quick action in this field will lead to competitive advantage in the struggle for new customers, and new sources of cash flow.

This change in the law, that facilitates patenting of new financial services that are implemented by software, takes place at the same time that other market forces are causing tremendous change in both the types of products provided in the financial marketplace, and the business entities and organizations that provide these services. This turmoil in the financial industry creates many opportunities that can be exploited by financial service patents.

Because all financial companies use software now, and because software is newly patentable where it provides new functions or old functions in a new way, we have developed the saying "every company is a software company now". A corollary to this is that "every company should exploit its software patent portfolio opportunities". It is clear that every financial institution or service provider should now act to protect its new financial products and software with patent strategies. Likewise, new products and services should be carefully reviewed before market introduction (in the planning phase), to avoid possible infringement of the patent portfolio of competitors.

The New Case

On July 23, 1998, the Federal Court of Appeals for the Federal Circuit, handed down its decision in the case of *State Street Bank & Trust Co. v. Signature Financial Group, Inc.*, 149 F.3rd 1368, 47 U.S.P.Q. 2d 1596 (Fed. Cir. 1998). This was an appeal of the 1996 lower court case from the Federal District Court for the District of Massachusetts, reported at 927 F. Supp. 502, and at 38 U.S.P.Q. 2d 1530.

In this case, Signature Financial Group owned U.S. Patent No. 5,193,056 for a computerized accounting system used to manage a particular type of mutual fund investment structure, specifically the "hub and spoke" structure for mutual funds that simultaneously invest in stocks in different national stock exchanges, priced in different currencies. The lower court issued a partial summary judgment that the patent was invalid and unenforceable because it claimed subject matter which was not patentable material. Specifically, said the lower court judge, the patent claimed software that was, practically speaking, essential for implementing the hub and spoke mutual fund structure. Therefore, the patent, in effect, monopolized a "business method". The Federal Circuit reversed this summary judgment, and held that the patent covered patentable subject matter and remanded the case to the District Court for a decision on conventional patent infringement fact questions, such as novelty and infringement.

The Winning Claims

As an example of the claims that were upheld as patentable subject matter, in the patent in question claim 1 read as follows:

1. A data processing system for managing a financial services configuration of a portfolio established as a partnership, each partner being one of a plurality of funds, comprising:

(a) computer processor means for processing data;

(b) storage means for storing data on a storage medium;

(c) first means for initializing the storage medium;

(d) second means for processing data regarding assets in the portfolio and each of the funds from a previous day and data regarding increases or decreases in each of the funds assets and for allocating the percentage share that each fund holds in the portfolio;

(e) third means for processing data regarding fairly incremental income, expenses, and net realized gain or loss for the portfolio and for allocating such data among each fund;

(f) fourth means for processing data regarding daily net unrealized gain or loss for the portfolio and for allocating such data among each fund; and

(g) fifth means for processing data regarding aggregate year-end income, expenses, and capital gain or loss for the portfolio and each of the funds.

The title of the patent, U.S. Patent No. 5,193,056, was "Data Processing System for Hub and Spoke Financial Services Configuration", issued March 9, 1993, naming R. Todd Boes as the inventor.

State Street and Signature Financial are both in the business of acting as custodians and custody agents for multi-tiered partnership fund financial structures, as targeted by the patent. State Street negotiated with Signature Financial for a license to use the patent and data processing system described in the patent. However, negotiations broke down and State Street brought a declaratory judgment action against Signature Financial asserting invalidity, unenforceability, and non-infringement of the patent.

The appeals court in *State Street Bank* said several things in the case that opened the flood gates for financial service patents, and added further fuel to the fire for the race to software patents in all industries.

Most importantly the appeals court indicated that the patent claims were directed to statutory subject matter. This is despite the fact that the claims cover a software system which may be claimed broadly enough to preempt all computerized bookkeeping for hub-and-spoke funds, in a context where computerized bookkeeping is absolutely essential for implementing this mutual fund structure. Indeed, the court says "it is essential that these calculations are quickly and accurately performed ..., given the complexity of the calculations a computer or equivalent device is virtually necessary to perform the task." In other words, the court acknowledges that by a patent monopoly on the software to implement the mutual fund structure, the patent owner may in effect monopolize that mutual fund structure itself. Yet this is no bar to patentability.

The court goes on to say that the patent claims are simply for a type of "machine", which is a conventional target for patents. (The Court implies that method claims could also have been used, although in this particular case, method claims were dropped from the prosecution of the application during the application phase).

The court also enters into an arcane patent lawyers' discussion about "means plus function" claims. The court states that claims citing a machine in the preamble having "means plus function clauses in the claim can be viewed as process claims, only if the supporting machine structure is not indicated in the written description for each means elements." Of course, the court later on seems to state that method claims are themselves adequate, so that "means plus function" claims without supporting structure may also be valid and patentable as method claims. The distinction between claiming a machine or claiming the method is probably of little interest to most clients in the software area, except for practicing patent attorneys.

Indeed, the court states that for purposes of patentability of subject matter "it is of little relevance whether the claim is directed to a machine or a process as long as at least one of the four enumerated categories of patentable subject matter [in Section 101 of the patent statute] is indicated, machine and process each being one such category."

Two Failed Attacks

The reasoning by which the court reaches this conclusion provides very broad encouragement to financial service patents and software patents of all types. Particularly the court addresses and rejects the two main traditional avenues to attack financial service patents specifically (which are a species of software patents), and software patents in general. These two exceptions are the so-called "mathematical algorithm exception" to patentability, and the "business method exception" to patentability.

The Death of the Mathematical Algorithm Exception

In discussing the mathematical algorithm exception, the court almost totally destroys this line of attack on software patents. The Federal Circuit acknowledges that "the Supreme Court has identified categories of subject matter that are unpatentable, mainly laws of nature, natural phenomena, and abstract ideas. The Supreme Court explained that certain types of mathematical subject matter standing alone represent nothing more than abstract ideas <u>until reduced to some type of practical application, i.e., a useful concrete and tangible result</u>" [emphasis added]. The Federal Circuit goes on to say that in the earlier case of *In re Alappat*, 33 F.3d 1526, 31 U.S.P.Q. 2d 1545 (Fed. Cir. 1994), we held that data transformed by a machine through a series of mathematical calculations to produce a smooth wave form display on a ... monitor, constituted a practical application of an abstract idea (a mathematical algorithm, formula, or calculation), because it produced a useful concrete and tangible result, the smooth wave form." That is, the court is saying that a display of a computer monitor is a "tangible result", and hence patentable subject matter.

Even more aggressively attacking the issue, the court went on to say that in the case of *Arrhythymia Research Technology, Inc. v. Corazonix Corporation*, 958 F.2d 1053, 22 U.S.P.Q. 2d 1033 (Fed. Cir. 1992), the Federal Circuit held "the transformation of electrocardiograph signals from a patient's heartbeat by a machine [a computer] through a series of mathematical calculations [by software] constitute a practical application of an abstract idea ... because it corresponded to a useful concrete or <u>tangible</u> thing --- the condition of a patient's heart" [emphasis added]. In other words, the prediction of the likelihood of a heart attack within the next thirty days, which was the result of the litigated patent, is "tangible" and hence patentable subject matter in the eyes of the appeals court. If this is "tangible", it would be hard to imagine anything that a computer does with software and data that does not result in a "tangible" result and hence qualify for patentability, if it otherwise novel.

Indeed, the court goes on to state that "today we hold that the transformation of data representing discreet dollar amounts by a machine [a computer programmed with software] through a series of mathematical calculations into a final share price constitutes a practical application of a mathematical algorithm, formula, or calculation because it produces a useful concrete and <u>tangible</u> result --- a final share price momentarily fixed for recording purposes and even accepted and relied upon by regulatory authorities and in subsequent trades." If calculating a share price for a trade is adequately "tangible" for patentability, as is a monitor display, then all functioning software must pass this "tangible" test. Indeed, practically speaking, this tangible test is now dead.

This reasoning is in keeping with the original case regarding patents for financial services, *Paine, Webber, Jackson & Curtis, Inc. v. Merrill, Lynch, Pierce, Fenner & Smith, Inc.*, 564 F. Supp. 1358, 218 U.S.P.Q. 212 (D. Del. 1983), regarding the CMA patent. This patent, of course, did nothing more than claim the software for an accounting system that produced a monthly statement for a particular type of account, such as the Merrill Lynch CMA account.

It is interesting that the Federal Circuit in *State Street Bank*, when discussing the Supreme Court case *Diamond* v. *Diehr*, 450 U.S. 175 (1981), states that the older, narrower requirement that software, in order to be patentable, should transform or reduce an article to a different physical state or thing, was <u>only an example</u> of the kind of function that was necessary for patentability and <u>not necessarily a requirement</u>. Of course, there is little left of this requirement for software to be patentable that it change something to a different physical state, if this "tangible" result is now interpreted to include transforming matter to a "yes or no" decision as to whether or not a patient may have a heart attack in the future (the *Arrhythmia* case), or indicating a share price that may be used for trading stock (as in the *State Street Bank* case, or in any of the program trading algorithm patents now being issued by the Patent Office).

The Federal Circuit, in *State Street Bank*, goes on to state "whether the patent claim encompasses statutory subject matter should focus not on which four categories of subject matter in Section 101 of the patent statute it is directed to (that is, process, machine, article of manufacture, or composition of matter) but rather on the essential characteristics of the subject matter, in particular its practical utility." (Of course, besides patentable subject matter, all patents must also have novelty, non-obviousness, and adequacy of disclosure and notice.) Summing up, the court says that claim 1 "is directed to a machine [computer] programmed with the hub and spoke software and admittedly produces a useful, concrete and tangible result. This renders it statutory subject matter even if the useful result is expressed in numbers such as a priced product, percentage, cost, or loss". This reading is so broad and so favorable for software patents in financial service industries, that it would be difficult to conceive of any software developed in the financial industry that would not be patentable subject matter, if it merely satisfied the minimal requirements for novelty, non-obviousness, notice, and disclosure. This would be the case even if the software were so broadly claimed as to monopolize all practical computer methods necessary for keeping the books and records for the product offering.

Hence, although the mathematical algorithm exception is not completely destroyed, it is so nearly destroyed by this case as to be little more than a drafting tip for the patent attorney for software and computer applications. In other words, the patent drafting should be clear enough to express that the algorithm in question is executed by a programmable device, and is not claimed so broadly that an individual thinking by himself in his office without any computers or software can infringe the claim merely by contemplation.

The Death of the Business Method Exception

The Federal Circuit is even more aggressive with the old business method exception, which has been used to attack financial service patents specifically, and software patents in general. This old judicially created exception to patentable subject matter states that although the patent statute allows for the patentability of methods without qualification, some judges have felt that this does not apply to "business methods", whatever that may mean. In *State Street Bank*, the court says "we take this opportunity to lay this ill conceived exception to rest.... Since the 1952 Patent Act, business methods have been and should have been subject to the same legal requirements for patentability as applied to any other process or method. The business method exception has never been invoked by this court or the CCPA [the Court of Patent Appeals, the predecessor of the Federal Circuit] to deem an invention unpatentable". The court goes on to favorably refer to Judge Newman's dissent in *In re Schrader*, 22 F.3rd 290, 30 U.S.P.Q. 2d 1455 (Fed. Cir. 1994), stating "[the business method exception] is an unwarranted encumbrance to the definition of statutory subject matter in Section 101, and should be discarded as error prone, redundant and obsolete. It merits retirement from the glossary of Section 101." With this case, the Federal Circuit has adopted the Newman view.

The Federal Circuit, in reviewing the lower court case that it reversed, notes that the lower court "noted as its primary reason for finding the patent invalid under the business method exception is as follows: "If Signature's invention were patentable any financial institution desirous of implementing a multi-tiered funding complex ... on the hub and spoke configuration would be required to seek Signature's permission from the inception of such a project. <u>This is so because the '056 patent is claimed [sic] sufficiently broadly to foreclose virtually any computer implemented accounting method necessary to manage this type of financial structure.</u>" [emphasis added] The lower court found this to be grounds for invalidating the patent. However, the Federal Circuit now finds that this "problem" is no problem at all, and is indeed an advantage deserved by the owner of the patent. The Federal Circuit states "whether the patent's claims are too broad to be patentable is not to be judged under Section 101 [this section defining patentable subject matter] but under other sections requiring novelty, non-obviousness, disclosure and notice, as with all patents. Assuming the above statement [by the District Court] to be true, it has nothing to do with whether what is claimed is statutory subject matter". The court went on to state "in view of this background it comes as no surprise that in the most recent edition of the Manual of Patent Examining Procedure ... a paragraph of Section 706.03(a) was deleted" that had been found in past editions.

The deleted portion read: "Though seemingly within this category of process or method, a method of doing business can be rejected as not being within the statutory classes." However, the Patent Office itself had rejected this old interpretation, even prior to the *State Street Bank* case. The court goes on to state "this acknowledged practice is buttressed by the U.S. Patent and Trademark Office 1996 Examination Guidelines for Computer Related Inventions, which now reads: Office personnel have had difficulty in properly treating claims directed to methods of doing business. <u>Claims should not be categorized as methods of doing business. Instead such claims should be treated as any other process claims</u> [emphasis added] ... We agree that this is precisely the manner in which this type of claim should be treated".

This case gave rise to the term "business method patent" to describe software enabled patents of all types. This term arose as this case became known and attorneys (many of whom were unfamiliar with the subject) attempted to analysis and explain it. Their thought process was that if the "business method doctrine" had been used to attack a particular type of patent, then that type of patent must be the "business method patent". However, the "business method doctrine" is actually an old doctrine that antedates all software, and was only relatively recently revivified to attack software patents. The term "business method patent" in fact is very new and did not appear until the "business method doctrine" itself was killed by this case. We find that better terms, such as e-patents, software patents, or service patents, often get more directly to the business import of what is protected. We disfavor the term "business method patent" simply because it often fails to practically communicate what it patents. Most people who know what software is, understandably do not know what a "business method" is. Indeed, the case law refers to the "business method" concept as "fuzzy". See *In re Schrader*, 22 F.3d 290, 30 U.S.P.Q.2d 1455 (Fed. Cir. 1994).

The term "business method", as commonly used in this book and generally in the business, should not be necessarily confused with a similar new statutory definition in the 1999 U.S. patent statute amendments. These amendments added a new section, 35 U.S.C. 273 (a)(3), which states that "the term "method" means a method of doing or conducting business". This is a particularly unenlightening part of a badly written bill of amendments, and it remains to be seen what this statutory definition will come to mean. However, nothing in this book referring to "business method" is intended to necessarily be included in this statutory definition, whatever it may mean.

(Note that it is possible to patent uncomputerized manual business methods, if they are novel, but this is at best a very small slice of the business method patent market.)

New Activity

Even before this case, a small number of innovative banks, insurance companies, and other financial players have seen the future and started to develop patent portfolios oriented around software enabled developments. We can expect more of this activity now. We can also expect more actions to enforce these patents, to obtain cash damages for infringement, to obtain permanent injunctions from infringers who continue to infringe, and to license patents to competitors. The injunctions would in effect have the role, with a good financial patent, of removing competitors from the product or service claimed by the patent, for the life of the patent. In the case of an injunction, this can have a dramatic effect on market share. By reducing competition, this may also increase profit margins. We can also expect increased licensing programs (which in themselves will provide independent sources of cash flow for the licensor) when a monopoly strategy is not pursued.

We can expect patents to continue to develop in the following areas: new types of financial services (such as new credit card account features), home banking, remote banking, e-money, e-commerce, Internet applications for banking, back office software (that increase administrative efficiency and responsiveness), and the design of new financial services and ways to deliver them (such as new types of mutual fund structures, new program trading algorithms, and new kinds of combined accounts and billing systems). Also, we will see new types of insurance products, annuities, insurance plans and products, and insurance coverage. The insurance industry will see more patents for software, eliminating personnel costs in the underwriting decision, and compressing the standard multi-tiered distribution system for insurance products (now including the insurance agent, the underwriter, and the reinsurer).

Early examples of patents in all these areas are now issued and publicly available. They are owned by institutions, individuals, consulting firms, and other vendors, in their respective industries. These patents are merely the innovative first rush to capture new intellectual property turf in a dynamic financial services industry. Tomorrow's winners of the innovation and market share battle (both from new technology on the Internet, e-money, and smart cards, and from the merger of financial institutions on the corporate side) will, to a greater extent than ever before in this industry, be impacted by successful patent strategies.

In the 1980s, Merrill Lynch broke new legal and business ground with its CMA patent. This allowed Merrill to stake out the rights to the CMA account and put Merrill in the status of a cash flowing patent licensor of its two CMA patents. It appears that Signature Financial has achieved this status with its hub and spoke patent. We can estimate that today there are many more such valuable financial patents recently issued or in the application process by financial institutions. These patents will bear fruit in the future with license royalties or monopolized patent positions in new and existing financial markets.

Culture of Confidentiality

As the financial services industry enters the patent world, it must implement the culture of confidentiality that is common in technology companies. Employment agreements, vendor contracts, and other documents must be revised to explicitly deal with ownership of intellectual property developed in the future. New services should be carefully reviewed for patent opportunities and infringement risks. Internal policies commensurate with the federal *Economic Espionage Act of 1996*, to maintain and respect trade secrets, should be initiated. (See Chapter 26 of *Patent Strategies for Business, third edition*, the companion volume to this book, for a detailed discussion of this topic.) And most, of all, business decisions need to be implemented to exploit the new opportunities for patent oriented cash flow, and market advantage, while maintaining the best patent defense posture.

3

e-Patents and Business Methods, The Beat Goes On: The *AT&T v. Excel* Case

A new case, *AT&T Corp. v. Excel Communications, Inc.* 172 F.3d 1352, 50 U.S.P.Q. 2d 1447 (Fed. Cir., April 14, 1999), further extends the pro-software e-patent ("business method") trend of the Federal Circuit from *State Street Bank & Trust Co. v. Signature Financial Group, Inc.* 149 F.3rd 1368, 47 U.S.P.Q. 2d 1596 (Fed. Cir. 1998), and earlier cases. (It is interesting to note that *State Street Bank* was handed down by Judges Rich, Plager and Bryson, and written by Rich, whereas *AT&T v. Excel* was handed down by Plager, Clevenger, and Rader, and written by Plager. This gives further confidence that this trend can not be ignored by a particular panel at the Federal Circuit, or reversed by an *en banc* decision.)

In *AT&T v. Excel*, the District Court held that AT&T's patent 5,333,184 was invalid under Section 101 for failure to claim statutory subject matter. However, the Federal Circuit on appeal found statutory subject matter and reversed and remanded for a review of patentability on novelty and other grounds.

The patent in question was for a "call message recording system for telephone systems". It described a message record that was used by a long distance carrier in providing differential billing treatment for subscribers. The independent Claim 1 referred to a "method for use in a telecommunication system" consisting of two steps. The District Court concluded that the amended claims of the patent implicitly recited a mathematical algorithm, and as such involved only the physical step of data gathering for the algorithm, and as such was per se unpatentable subject matter.

On appeal the Federal Circuit found that no "physical transformation" is necessary for a mathematical algorithm to be patentable subject matter.

The Federal Circuit went on to state that "we consider the scope of Section 101 [the definition of patentable subject matter in the statute] to be the same regardless of the form -- machine or process -- in which the particular claim is drafted ... In fact, whether the invention is a process or a machine is irrelevant."

The Federal Circuit found that the claimed method of adding a PIC [Primary Interchange Carrier] indicator to a message record for a long distance telephone call in a billing system was "a useful non-abstract result that facilitates differential billing of long distance telephone calls... Because the claimed process applies the Boolean principle to produce a useful, concrete, tangible result without preempting other uses of the mathematical principal, on its face the claimed process comfortably falls within the scope of Section 101 [patentable subject matter]." The "tangible" result was "providing differential billing treatment for subscribers depending on whether a subscriber calls someone with the same or different long distance carrier." Favorably citing *Arrhythmia Research Technology, Inc. v. Corazonix Corporation*, 958 F.2d 1053, 22 U.S.P.Q. 2d 1033 (Fed. Cir. 1992), the court said "that the product is numerical is not a criterion for whether the claim is directed to statutory subject matter".

e-Patent Strategies

The court went on to state that "the notion of physical transformation can be misunderstood. In the first place it is not an invariable requirement [of patentability] but merely one example of how mathematical algorithms may bring about a useful application... [as] an example [it is] not an exclusive requirement." The court also favorably cited the *State Street Bank* case saying "after Diehr and Chakrabarty, the Freeman-Walter-Abele test has little if any applicability to determining the presence of statutory subject matter ... The mere fact that a claimed invention involves inputting numbers, calculating numbers, outputting numbers, and storing numbers, in and of itself, would not render it non-statutory subject matter, unless, of course, its operation does not produce a useful, concrete and tangible result".

Further the court stated "it is now clear that computer based programming constitutes patentable subject matter so long as the basic requirements of Section 101 are met [novelty and utility]."

AT&T v. Excel makes it clear that novel software based inventions are as patentable as any other technology. For novel services that may only be delivered by software, such services can effectively be patented by patenting the software, in the general sense, that is providing these services.

The European Patent Office

The pro-software patent trend in the United States is also followed by a decision of the Board of Appeals of the European Patent Office, handed down February 4, 1999, regarding the Applicant, IBM Corp. In this case, IBM appealed claims that were rejected by the EPO for a "method and system in a data processing system windowing environment for displaying previously obscured information". Claims 1-6 involved preambles for methods and computer systems for executing the indicated method, and novelty and inventive step were found by the EPO. However Claim 7 and subsequent claims were directed to a computer program stored on a computer readable storage medium. Specifically "a computer program product comprising a computer readable medium having thereon: a computer program code means, when said program is loaded, to make the computer execute procedure" which would execute the method claimed in the previous claims.

The Board found this language to be patentable. To do this, the Board went through heroic efforts of rationalization noting that the Guidelines for Examination of the EPO states "a computer program claimed by itself or as a record on a carrier is not patentable irrespective of its content", and further noting that Article 52(2)(c) of the EPC (*European Patent Convention*) states that "programs for computers shall not be regarded as inventions... and are therefor excluded from patentability". To find patentability, the Board noted that Article 52(c) of the EPC "establishes an important limitation to the scope of this exclusion. ..." According to this provision, the exclusion applies only to the extent to which a European patent application... relates to programs for computers as such". The Board goes through a heroic interpretation of "as such" and determines that software programs excluded from patentability are those that are "mere abstract creations lacking in technical character", and that therefore all programs for computers "must be considered as patentable inventions when they have a technical character". Although computer programs "cannot per se constitute technical character avoiding the exclusion", this technical character can be found if "the software [is used] to solve a technical problem."

The Board remanded the application to consideration by the European Patent Office "taking into account the fact that a computer program product is not... excluded under all circumstances".

This decision allows the claimed format of software on a disk executing a novel process, that is commonly used in the United States today.

e-Patent Strategies

This claim format allows us to pursue direct infringement against a software pirate which copies the software on disk, but never actually executes the software on a machine. This allows direct infringement litigation against such a pirate, without the requirement of suing a customer who actually uses the pirated software, and then pursuing the pirate in a contributory infringement derivative action.

We can expect further cases to develop in the same main stream pro-software e-patent direction that has been firmly established by the Federal Circuit in *State Street Bank* and *AT&T v. Excel*. The frenzy for "business method" patents will continue to grow, with this judicial support.

4

e-Patents for Internet and Banking Services: The Survey

On January 4, 1999, we updated our survey of U.S. banking e-patents and U.S. Internet e-patents. These are, generally speaking, a species of software patents, that is, "business method" patents. The primary finding of the survey is that the numbers of both banking and Internet e-patents are growing at an exponential rate. In the case of Internet patents, the explosive growth exceeded even the estimates that resulted from our June 1997 survey. (See Chapter 30 of *Patent Strategies for Business, third edition*, the companion volume to this book).

Banks as Owners

390 patents were owned of record by banks. A large number of these were clearly unrelated to banking and probably involve holding title to patents as security for loans and other financial obligations. Of the patents with banking related substance matter, a fair number related to equipment for processing bills and coins.

A large minority of the bank owned patents covered software systems for bank services and systems. These included credit reports, smart card security, the monitoring of fund floats, checking systems, document image storage recognition and analysis, electronic funds transfer instruments, holographic check authentication, fail safe on-line financial systems, automatic teller machine systems, anti-counterfeiting laminated currency, remote banking, home banking, smart cards, and others.

Banking Subject Matter

Further, it is noted that the number of patents that dealt with subject matter involving computer applications in banking, regardless of ownership, was 510. This with the above information would indicate that a wide variety of computer and software applications for banking are not owned by the banks that are constrained by them. This may indicate that banks should become more aggressive in owning patents for the new products and systems that they themselves develop and use, rather than let others (particularly vendors and consultants) monopolize this market and its profits.

Banking and the Internet

Also, it is particularly interesting that the number of patents involving banking and the Internet, regardless of ownership, numbered 22. However, the first of these was not issued until November 25, 1997. This entire patent field has started from scratch and has grown rapidly within the last year. It is anecdotally interesting that the first such banking Internet patent was issued to Mastercard International, Inc. We can expect that many more applications of this type are pending.

Mastercard

Mastercard is indicated as the owner of record of 10 U.S. patents, 4 of which it has received in the last year. Some of these patents involve software systems for cash free and remote electronic transactions. The others involve patents for devices interfacing with such systems.

Citibank

Citibank or Citicorp are indicated as the owners of 42 patents of record. It is interesting that all of these patents are related to banking, and none appear to be loan collateral. This indicates that Citicorp does not use a system of taking absolute assignments as security in intellectual property, but may be using the preferred system of chattel mortgages or even the more doubtful system of the UCC regime alone.

The number of Citicorp patents has doubled since 1991, that is, within the last seven years. However, the majority of the oldest 50% of the Citicorp patents are design patents and not utility patents. However, all of the newest 50% of the Citicorp patents, except one, are utility patents and not design patents. This indicates a major commitment by Citicorp within the last seven years to patent its proprietary software for e-commerce, remote banking, and software systems to support its operations.

Home Banking

33 patents are indicated for home banking or remote banking subject matter. The number of these patents has doubled since March 1997, that is, within the last year and one-half. This indicates explosive growth in this area. All the home banking and remote banking patents refer to software and computer systems to implement the same.

The Internet

609 patents involved Internet applications. It is a stunning statistic that of this number, half the patents have been issued since

May 1998, that is within the last half year. This explosive acceleration of growth in the number of Internet patents indicates an even bigger number of Internet patent applications in the pipeline at the Patent Office. This statistic also indicates a pending patent shootout over market share for new developing areas of Internet based applications.

Smart Cards

322 patents are indicated for smart card subject matter.

It is a striking statistic that of these smart card patents, half of them have been issued since March 1997, that is, since less than two years ago. This indicates an explosive growth in the number of smart card patents, and guarantees that a even larger number are pending in the pipeline at the Patent Office.

The Losers' Response

The two leading cases establishing patentability for software patents for financial services are *Paine, Webber, Jackson & Curtis, Inc. v. Merrill, Lynch, Pierce, Fenner & Smith, Inc.* 564 F.Supp. 1358, 218 U.S.P.Q. 212 (D. Del. 1983), and *State Street Bank & Trust Co.* v. *Signature Financial Group, Inc.* 149 F.3d 1368, 47 U.S.P.Q. 2d 1596 (Fed. Cir. 1998). In both these cases, the losers bet against financial patents, and lost money when the patents were supported. In this light, it is interesting that both the losers, Paine, Webber and State Street Bank, are yet to be indicated as owning any patents. (Some dogs appear to learn new tricks slower than others.) This is despite the fact that observers to these cases, such as Citibank and Mastercard, have already committed themselves to the new opportunities available in financial patents.

Specific Industrial Intelligence

For a particular company, a more specific and detailed analysis can be executed for the patent developments in their field, both by subject matter and according to ownership by their key competitors. This would indicate the direction of the competitors' patent portfolio from which they can expect offensive action. This can also suggest holes in the investigating company's own patent development that should be filled with patenting activities and perhaps product development.

The analysis of this survey covered only the U.S. issued patent database. A more detailed specific company analysis might also include published foreign patent applications (U.S. patent applications are not published). Covering foreign published applications might give an advanced indication of patents that may appear on the scene in the U. S. at a later date, since many foreign applications are also filed in the U.S.

5

Rules of Virtual Genius: Software and Internet Update

"I don't invent anything, until <u>after</u> I find a customer to buy it."

-Thomas Edison

The rise of software patents (that is, e-patents, or business method patents) and their application to Internet services, software applications of all sorts, insurance and financial services, virtual retailing, and business plans, has led to an addition and update of the rules of virtual genius (about how to invent in a corporate environment), described in Chapters 2 and 3 of *Patent Strategies for Business, third edition*, the earlier companion volume of this book.

The following are the new additional rules for software and Internet invention, for profitable corporate activities.

Rule 13: Internet-ify.

Apply the Internet and patent your applications. Develop new services enabled by the Internet, or provide new platforms that use the

Internet to enable your old established services that were not previously provided by the Internet.

When you take these steps you will likely find opportunities to deny your competition the right to copy your applications, or to reverse engineer them, during the term of any patent for your advances. The patent opportunities may extend to the software application, the service provided, the hardware and software platform that facilitates the service, or the GUI (Graphical User Interface) that is supported by the system.

A fertile area for this rule is telecommunication services and remote services. For example, where previously a leased private telephone line may have been used as part of a system architecture, for example for a wide area network, replace that private telecom line with the Internet. This indeed has been the essence of the rise of VPNs (virtual private networks) that have replaced many older WAN architectures.

Rule 11: For Software Only -- 1999 Expansions.

Rule 11 was first discussed in Chapter 2 of *Patent Strategies for Business, third edition*, the companion volume of this book. Rule 11 is expanded here.

Rule 11a: Find New Functions.

Anytime you develop a new function that is facilitated by software, evaluate the possibility of patenting the software system delivering that function, to prevent your competitors from reverse engineering and competing with you in supplying that function. This is true for any software system. (See for example, the Stac Electronics Inc. Stacker data compression product, or spreadsheet software, or virtual bookstores on the Internet). A particularly fertile area to find new patentable functions are in the classic service industries such as telecom services, financial services, insurance, retail sales, and the newer Internet services.

Rule 11b: Assemble New Combinations of Old Functions, Provided by Software.

It is an old principle of patent law, which was originally developed for the mechanical arts, that a "new combination of old parts" can be patentable in itself. One example of this may be the Wright brothers airplane. This was a mechanical invention that basically assembled a pre-existing German glider design, with the internal combustion engine (which was not developed by the Wright brothers), with a propeller or "air screw" (which was not developed by the Wright brothers, but was refined by the Wright brothers for the airplane). The resulting new combination with mostly old parts resulted in a fundamentally new invention with unprecedented results, that is, a heavier than air flying machine.

Now with the advent of software inventions, we have an analogous role for the "new combination of old functions". These are sometimes called "Swiss army knife" inventions, since, like a Swiss army knife, they endeavor to find a new level of utility by creatively assembling a variety of different old functions or pieces into one product or platform. In the software area, it can be very useful and very "non-obvious" to invent efficient ways to bundle various functions that previously have been executed by incompatible software platforms, into one integrated system sharing common compatible databases. Facilitating this combination of functions often leads to a superior level of service, with reduced costs, and can represent an inventive step forward for society.

Rule 11c: Provide Old Software Functions with New Hardware/Software Infrastructures.

Providing old functions and services with new and superior hardware/software platforms can provide great business opportunities and opportunities for patents. For example, see the replacement of pre-existing cellular analog phone systems by digital PCS mobile phones. Also, see the replacement of wide area networks on telephone leased lines by virtual private networks using encrypted communication over public and "free" Internet lines.

Rule 14: Develop a User Friendly (Intuitive) GUI, Patent It, and Make Sure it Downloads Quickly.

Innovative GUIs (Graphical User Interfaces) can be patented as virtual machines, much like the actual dedicated black boxes of an earlier generation were patented as actual machines or apparatuses. An intuitive, good looking, and fast downloading GUI can be an extremely valuable portal to any hardware/software system that provides useful services. As such, these GUIs can be worth patenting, where possible.

Likewise, new GUIs can offer opportunities for trademarks, copyrights, and design patents, which can all act together to help suppress copying of distinctive GUIs by competitors.

Rule 12a: Mind the Aesthetics.

Even in software and computers, minding the aesthetics is an important rule. Rule 14 discussed herein is partially an aesthetic application to GUIs.

We have long advocated interesting aesthetic design for the PC industry. Particularly as the PC industry moves into "commodity" status, it will be a particularly useful if PCs junk their beige square boxes in favor of "packages" designed by, for example, any good Italian design firm in Milan. Examples of this approach most recently have been the Apple iMac, new Apple Macintoshes, and some Sun Microsystems products. This has brought both color and shape to the PC industry, finally, after two decades.

It typifies this trend, that Steven Jobs' first big success was creating the PC at Apple (it was an ugly little product that sold itself on function and price), while Mr. Job's latest success has been resurrecting Apple by putting the product in a pretty package (although it does the same old thing).

Swatch watches and Swatch cars have also brought aesthetics to products of industrial technology. The Swatch car uses interesting new design and manufacturing techniques to enable its eye catching design. Swatch watches, at least in the Swatch "Skin" line of watches, uses some modest new technology to make an eye catching fashion statement that is well evaluated in the marketplace.

Rule 15 (extension of Rule 5 -- Apply New Devices): Apply New Software Functions and Protocols.

New functions are available from new software and they should be applied to your business wherever useful. These new applications to your business will, in turn, present opportunities for patenting the resulting new competitive advantages, in order to keep these advantages away from competitors. These new software functions include public key encryption, voice recognition, voice to text, text to voice, and hands-free interfaces facilitated by voice to text, to name a few.

Also, new protocols are allowing new levels of service at new lower price structures. These include the Internet 2 protocols, voice over IP, fax over IP, reserved bandwidth, real time IP (streaming format protocols), and VPNs (virtual private networks). These developments seem to apply as well to intranets as to the Internet.

Applications of extranets may also provide patentable competitive advantages. (Here we use extranet as a private system with a TCP/IP protocol, or a virtual private network, with some restricted partial level of public access). For example, this may include the ability of the public to access a web page and order products, in a system that will directly access and modify in-house inventory, shipping and receiving databases as a result of the order.

Rule 16 (an extension of Rule 5a - Apply New Devices): Apply New Digital Telecom Devices.

PDAs (personal digital assistants) palm top computers, smart mobile phones (for example, with encryption and/or GPS), pagers with visual displays, PC digital cameras for the Internet, together with streaming video and audio over IP, cable modems, and PCS mobile phones (digital mobile phones), all are interesting new hardware features at reduced prices that have recently become realistic to incorporate into patentable hardware and software systems to deliver new services, or to deliver old services in new ways at new reduced prices. Consequently, these new applications present patentable opportunities.

Rule 17: Use Extranets and Artificial Intelligence to Eliminate Distribution Channels and Costs for Goods and Services.

We have seen applications of this rule to books with Amazon.com, and to drugs with the affiliated Drugstore.com. We have also seen it in industrial hardware and in computers. See for example Dell Computer's web sales. We have seen it with financial services with Schwab and Fidelity's stock trading over the Internet. We have seen it with IPOs (a former principal of Hambrecht & Quist recently began to underwrite IPOs on the Internet direct to retail investors in an auction system). We have seen it with goods and services of all kinds in Priceline.com. We have just begun to see it with life insurance (see the Schwab/Kemper life insurance quotes on the Internet). We have certainly seen it with the distribution of investment information and analysis for stocks and bonds over the Internet in a variety of locations (containing sophisticated charting and quantitative screening, and distribution of analytical reports). We have seen it with the new use of extranets which let vendors and suppliers into your transactional databases through the Internet and into your software for sales orders, accounts receivable, shipping and receiving and databases for the same.

Many of these developments when they were new offered opportunities for patented advantages in the marketplace. Some of the innovators were nimble enough to apply for patents (see Priceline.com) and some apparently were not (see Amazon.com).

For example, when the Internet bookstore idea was new, it probably presented a patentable opportunity, although apparently Amazon.com did not pursue this patent. If Amazon.com had obtained a patent for the virtual bookstore, Amazon.com would probably not now be in a brutal fight for market share (that is killing its profit margins) with BarnesandNoble.com.

On the other hand, we see Priceline.com obtained a patent on its business before its public roll out. (Although this patent may be litigated soon, and Priceline.com appears to have copycat competition, its eventual market share may be profoundly influenced by the success or failure of its patent strategy. And Priceline.com at least now has a shot at enforcing a patent homerun. Amazon.com never gave itself even a shot at a patent homerun, apparently.)

As distribution systems continue to be fundamentally reorganized by these Internet software systems (in industries such as insurance, financial services, and retail sales), the first player to

Rules of Virtual Genius: Software

implement new methods of increasing distribution and lowering costs should take any opportunities available to suppress copycat competition, to boost its market share and to maintain its margins over the long run. For these new Internet retailing applications ("e-tailing"), it remains to be seen where the profitable business model lies in the new compressed Internet distribution systems, without patent protection to eliminate profit margin competition.

My favorite example of Internet applications to retailing, which may not have been patentable at any time, is Fromages.com. This single web page has allowed one cheese shop somewhere in France to distribute its products by air courier throughout the world. This is particularly interesting to the U.S. cheese market, where unpasteurized French cheese processes cannot legally be used to manufacture cheeses, and where unpasteurized French cheeses can be imported only for personal consumption and not for resale. Only by direct retailing from France can these extraordinary cheeses be eaten in America.

Rule 18: Incorporate Remote Payment.

Incorporating remote payment into your software-based remote sales system, whether it is by the Internet or other telecommunication means, provides opportunities to obtain a legitimate patent monopoly in new market areas. These opportunities include all areas of e-commerce, web credit card transaction authorization, smart cards, e-money, remote checking, and e-billing. Both these services and the software/hardware infrastructure for providing the services may represent patentable opportunities. Incorporating these methods into any business system can be opportunities of tremendous advantage.

Rule 19: Develop New Algorithms.

Algorithms for software based decision making can be patentable in their own right. This could have a tremendous economic impact for a company that would be worth protecting. Older smart equipment and software, although it may function successfully, is today in retrospect not always as smart as it could be. Developing new decision making algorithms, whether they are quantitative and specific, or heuristic, or based on neural nets or other decision making algorithms, can dramatically improve the function of decision based software and smart equipment. This also applies to pattern recognition, image processing, and encryption software. It is also quite effective for the patent portfolio for the RSA Corporation, and their dual public key/private key encryption/decryption algorithms. It has also been used for program trading and other quantitative decision making. It has also been applied to a variety of smart equipment and expert systems by the telecom and defense industries; for example, see GPS based encryption algorithms, and the conversion of smart bombs from a laser guided targeting algorithm (which is susceptible to line of sight weather conditions) to GPS based target navigation with predetermined target coordinates (which works better in no visibility situations).

The Antidote to Inventing Around:
Rule 20: Large Disclosure, Continued Prosecution, Late Claiming.

A good procedure to use to inhibit competitors from inventing around your patents, centers on the idea of keeping prosecution of one of your early patent applications open with continuations and continuations-in-part (CIP's).

In this procedure, an initial early application is filed with the most extensive disclosure possible at the time. Additional matter is added in CIP's as developed. When claims are allowed, they are issued, but the application is kept alive with other claims, in continuations. When competitors appear, their products are targeted as well as possible with any available new "late claims" that are enabled by the original specification or early CIP's.

Rules of Virtual Genius: Software

This is basically the Tollgate Strategy combined with the New Submarine Strategy. It is limited by the scope of the original enabling disclosure, and any early CIP's; but, it can ease the demands of completely inventing around yourself at the inception of your project, as an inoculation against your competitors inventing around you. (See Chapter 3 of *Patent Strategies for Business, third edition*, the earlier companion volume to this book, for a discussion of the antidote to inventing around, the Tollgate Strategy, and the New Submarine Strategy.)

6

Four Stages of Patent Denial in the Software Industry

Over the last few years, software patents have developed in the United States and their success in helping the bottom line for businesses that own software patents has been proven. This has caused many players in the software industry to go through levels of denial regarding software patents.

Stage 1: It's evil.

Originally we saw many software professionals that were offended by the idea of software patents because of some sort of anarchistic or libertarian ideological impulse against change and development in this area. Basically, they thought that software patents simply "should not" be allowed. In some sense, software patents were seen as morally wrong.

This issue has now been decided by our society and software patents are here to stay. So the question now has become one not of what should be done for the good of society, but what must be done for the individual business to accommodate the circumstances as we find them. (The economic and social arguments in favor of patents in general apply to all new technologies, including software. However, this issue seems to be the subject of a fight we go through every time a fundamentally new technology is developed.)

Stage 2: It's the Market Share, Stupid.

As it has become clear that software patents are here to stay, and whoever uses them has an opportunity to make a tremendous amount of money, the level of denial in the software industry against software patents has become more sophisticated, as it has eroded. As some software companies and owners of software applications are racing to the Patent Office to develop these software patent portfolios, others continue to hold back and be left behind with the argument that "it is not software patents that are important, it's the race to market share". That is, when a new product comes out, it is more important (the deniers say) to obtain market share; and market share is determined more quickly than it is possible to get patents out of the Patent Office. Then, by the time the patents are obtained, the original prior art is obsolete in the marketplace, the deniers assert.

There are several fallacies to this view. First, software patent applications can benefit from expedited review at the Patent Office and an accelerated issuance of patents.

The second and larger fallacy is that the determination of market share is, in the first years of a product, determined by market factors and probably not largely impacted by a portfolio of pending patent applications. However, the final determination of market share can be catastrophically, or triumphantly (depending on which side you are on) determined by a patent fight. This is because one of the results of patent infringement is not just an award for monetary damages for past infringement, but a permanent injunction for the life of the patent against the infringing party to no longer engage in the infringement. That is, the infringer may be put out of business entirely for the life of the patent. An example of this is the case of the *Stac Electronics v. Microsoft Corp.* (D.C. C.Cal CV-93-413-ER, May 13, 1994, June 8,

1994), about the original data compression product (which is discussed in more detail in *Patent Strategies for Business, third edition*, the companion volume to this book). In this case, Microsoft was found to infringe two software patents by Stac Electronics and as a result Microsoft suffered a permanent injunction. This may have reduced Microsoft's share of the data compression software market to 0%, regardless of whatever market share Microsoft had achieved by business means prior to the injunction. Perhaps, the only reason Microsoft Windows has a data compression feature today is that after the court defeat, Microsoft settled with Stac Electronics for an extremely expensive license to Stac's data compression software algorithms.

Hence, in the early stages of the development of a new industry, the race to market share is important and determined by business factors. However, once the market is developed, there is usually a shake out due to market forces. Then, the final market share allocation and market participant list is often determined by a patent fight. The essential lesson here is to position yourself to be a player in the eventual patent fight for market share. To do this, a company must very early in the development of the industry file its patent applications and obtain its early priority dates for those applications.

This scenario or general structure of development is a classic one for all new industries. It has been followed for data compression software in the *Stac Electronics v. Microsoft Corp.* litigation. It was also, in earlier days, followed for the wireless radio (Edison versus Marconi), for the airplane (the Wright brothers versus Glenn Curtis), for the automobile (Henry Ford versus everybody else), and for a variety of other industries created by Thomas Edison (who spent as much time in patent litigation as he did in the laboratory). Edison was quite successful in this endeavor and eventually ended up with significant earnings from major players in several industries that he initiated, including recorded music, motion pictures, radio, and electric power.

Stages of Patent Denial

We are at the front end of seeing the same sort of patent struggle being played out in the final resolution of market share and profit margins in new software enabled markets. The litigation is just starting, but the major patent positions are currently being staked out with patent applications with today's priority dates.

Stage 3: It's Already Obsolete.

A third state of denial for the software industry is that by the time the patents are obtained the products are obsolete. What this actually means is that you may start your market with release 1.0, but by the time the patent issues, release 2.0 rules the market. The fallacy of this thought is that any patent filed immediately prior to the roll out of release 1.0, probably also controls the release 2.0 product. So, although release 1.0 may be obsolete in the marketplace by the time release 2.0 rolls out, the patent describing release 1.0 may still describe the fundamental core functions and software engine of release 2.0, and that first early stage patent may very well still prevent any effective competition to release 2.0. That is, although release 1.0 may still be obsolete, the patent for release 1.0 may still rule and dominate later improvement releases of the product in the marketplace. To put it another way, release 2.0 is probably a bells and whistles refinement to release 1.0, and these bells and whistles and improvements would be meaningless without the fundamental and patented inner core product. New functions of release 2.0 would be as meaningless without the underlying functions of release 1.0, as the top layer of the wedding cake would be without the underlying three layers of cake (it would look like Ken and Barbie standing on a cupcake).

Also note that the subsequent second generation improvements of release 2.0 can also be patented in their own right with second generation add on patents.

Stage 4: We'll Be Out by Then.

Perhaps the most advanced stage of patent denial comes from some investors. Some short term investors, with an exit strategy to be out of the investment quickly, figure that by the time the patent shake-out comes to their market niche, they will have sold their position in the company. Therefore, they reason, it does not matter to them what the patent story is. The fallacy of this is that their sale price will in fact be profoundly influenced by the patent positioning of the company at the time of their sale. The value of Stac Electronics would have been about zero, if Stac had no patent to defend itself against Microsoft. But instead, Stac was evaluated at about $100 million, apparently, after it enforced its software patent (for an algorithm) against Microsoft. David can beat Goliath, and make a fortune, if David has a good patent in his sling.

7

Intellectual Property Audits: Start with the Business Plan

Special Tips for Software, Telecom Services, and Financial Services

Elsewhere in the companion volume to this book, *Patent Strategies for Business, third edition,* we discuss in Chapter 1 an intellectual property strategic management program. We present here some additional ideas regarding intellectual property audits, particularly for software, telecom services and financial services. Note in particular that in these three fields, new services and new infrastructures for providing old services, are now candidates for patent protection. These developments are also candidates for infringement of the prior patents of others.

Start with the Business Plan

Intellectual property audits should all start with your business plan. Only with knowledge of how your company plans to make money, can patent opportunities and threats be selected (or rejected) on the basis of what will most help your profits. This business driven approach is most clearly promoted by market-driven patenting, as discussed below.

Culture of Confidentiality

Probably the biggest change in software enabled service industries that are now entering the world of patents is the implementation of what I call a "culture of confidentiality". In particular, this requires all new product and service developments to be maintained as confidential trade secrets prior to their public roll outs. Also, the culture of confidentiality requires that any patent application be filed prior to the public disclosure of the trade secret. In addition, this culture of confidentiality requires a prophylactic title (ownership) regime as described below.

Prophylactic Title (Ownership) Regime

Contractual provisions and procedures must be put in place to avoid the pandemic title (ownership) defects that are common in the world of intellectual property today. Particularly, this requires all employees, consultants, vendors, potential customers, and other recipients of confidential disclosure, prior to the disclosure or their participation in the development of trade secrets and other intellectual property, to sign written agreements assigning all rights to the developments to the proper corporate entity.

These contracts are particularly easy to get many parties to sign prior to the development of the intellectual property in question, but can be quite difficult to arrange after such development has happened.

Also, note that normal business activity (absent these contractual agreements to cause title to come to rest in the proper place), may lead to title coming to rest in unintended and incorrect hands, with disastrous consequences to the proper corporate owner. Specifically, if patent applications are not filed early, then normal business activities to commercialize new developments may cause the ownership of the patent rights to the development to inadvertently lapse into the public domain, or worse yet, to unintentionally fall into the hands of an individual inventor on an exclusive basis.

Market Driven Patenting

The market driven approach to possible patenting requires that each new good or service that is contemplated be reviewed for patentability prior to its disclosure and roll out to the public. The patentability of software is not commonly understood by business and technical people in the software businesses, and is best appraised by a patent attorney familiar with software based patents.

Also, prior to rolling out new products or services, defensive patent analysis should be performed to ensure that the party developing the good or service has the right to sell the good or service without infringing on the prior patent rights of a competitor.

This is referred to as a "market driven" patenting approach because the protection of a previously defined market drives the push for patent weapons.

A larger view of market driven patenting involves an inspection of the business plan and methods by which the corporate client makes money. Then a creative search is made of possible patent opportunities to legitimately suppress competition and maintain profit margins for this plan. How this relates to planned new product roll outs is discussed above.

Regarding existing products and niches, the opportunity to patent the existing products may have passed if the products are already rolled out to the public. However, those old product niches may be placed into patent protection by the intentional development of second generation patentable improvements that will obsolesce prior non-patented products and services.

Sometimes, a good source of market driven patent ideas is the Marketing and Sales Department.

Technology Driven Patenting

The technology driven patenting process involves looking at the best new in-house ideas for products and services, whether or not they may be presented by the corporation to the marketplace, and protecting them with any available patent strategies. These patents may then be used to protect any eventual product roll out, or may be sold as assets to non-competing third parties if the corporation decides not to exploit these products and services itself.

This is called technology driven patenting because interesting technology (independent of any apparent markets) drives the push for patent property.

Legally Driven Patenting

Legally driven patenting may involve an inspection of the competition's products and services and their intellectual property position. Gaps in the competition's intellectual property protection may then be identified. The inspecting corporation may then invent into those patent gaps of the competition to attack the competition's products and markets with the inspecting corporation's own new patentable product improvements.

A second approach to legally driven patenting involves, first, research directed at the patented technology developments of the entire industry, conceptualization of the directions of development, and inventing into the future destination of industry development, possibly using a "submarine patent claims" approach and late claiming within 20 years of any original patent application date.

A third legally driven patenting approach would incorporate identification of a specific problem patent of a competitor that is blocking entry into a target product niche (i.e., identifying a potential patent infringement plaintiff/competitor), and then responding to the problem by "inventing around" the specific patent in question. This may allow entry into the product niche and attack on the target clients of the competitor, without exposure to infringement liability.

As a practical matter, a problem patent may identify itself. This can happen, for example, when (1) a company is sued for patent infringement, (2) a company receives a demand letter, a cease and desist letter, or an offer to take a patent license (implying litigation in the alternative), or (3) a company is paying expensive royalties on an existing patent license.

Being sued for patent infringement is probably the most urgent incentive to legally-driven patenting, but inventing around is rarely actually done in response to litigation. This is because it is difficult to invent on demand, and because the patent bar traditionally does do value based compensation, instead of hourly compensation. Defendants would invent around plaintiffs' patents much more often, if only they could. However, in a few cases, inventing around plaintiff's patents has, when done, yielded huge returns by avoiding plaintiff's permanent injunction, while remaining in the marketplace.

We call this approach "legally-driven" patenting because it is driven by a need to improve the existing legal environment presented by the patent environment that a corporation must work in. This is an underdeveloped approach to patenting, but can give high returns. It has unusual demands for inventive patent lawyering, working closely with business and technical people.

Bennchmarking

Another interesting aspect of intellectual property audits is benchmarking your corporate support of an intellectual property plan. This may involve obtaining or commissioning a survey of your industry to look at key management parameters and benchmarks for investment in patent portfolios, and for purchasing, hiring and pay scales for individuals within your organization supporting the patent portfolio development function.

Patent Profits

It is interesting that an increasing number of corporations in the U.S. have managed to make their patent portfolios become cash flowing profit centers. These may include, for example, Texas Instruments, IBM, Medtronic, Alza, and a variety of new privately-held start-up companies.

What used to be a headquarters staff overhead cost, that is patenting and licensing, is now budgeted as a profit center in more corporations. Although profits alone should not be the only goal of that function (some patents are best not licensed), positive cash flow certainly makes it easier to fund this operation in the annual budget.

8

Intellectual Property Audits: Special Steps

The following is a list of specific steps that may be executed in an intellectual property audit. An intellectual property program may be developed with all these steps or selected steps that are most applicable to a specific company.

Industrial Intelligence and Response

An intellectual property audit may be used for legitimate industrial intelligence to determine what your major competitors are doing and to respond to the same before severe problems are generated for you by your competitors. The following steps may be followed:

Step 1: List your major competitors.

Step 2: Inventory the current U.S. and foreign patents issued to your major competitors. Also, search the published foreign patent applications of your competitors that have not yet resulted in patents in the United States.

Step 3: Group the resulting inventory of competitor patents by product line and industry.

Patent Audits: Special Steps

Step 4: Characterize market niches and directions of the patent portfolios developing with your competitors.

Step 5: Determine if "invent-around" opportunities exist for you for any particular patent grouping of your competitors. (Rules for inventing-around competitor patents are discussed elsewhere in Chapters 2 and 3 of *Patent Strategies for Business, third edition*, the earlier companion volume of this book.)

Step 6: Determine whether you can use a leap frog and tollgate strategy for any group of competitor patents. (The leap frog strategy and tollgate strategy are discussed elsewhere in Chapters 2 and 3 of *Patent Strategies for Business, third edition*, the earlier companion volume of this book.)

Step 7: Determine if any group of competitor patents blocks any of your planned products or lines of business. If so, develop an invent-around strategy for each of your blocked future products.

Step 8: Plot the number of patents issued per year for each competitor and product grouping. From this determine the rate of growth of competitor patents in your industry. Use this percentage growth as a benchmark for your development of a patent portfolio for your own company.

Step 9: Develop a list of key terms and key fields for patent developments in your industry.

Step 10: From the list in Step 9, find the key patents in your industry, and determine who owns these key patents. Determine invent-around strategy possibilities for such patents.

A Market Driven Intellectual Property Audit for <u>Existing</u> Products and Services

A market driven intellectual property audit process for existing lines of business may have the following steps.

Step 1: Inventory all the intellectual property in your company at this time. This includes patents, trademarks, copyrights, and trade secrets. The inventory should also include license agreements, joint development agreements, partnership agreements and other contractual arrangements that may impact intellectual property. These may include bringing intellectual property into your company, transferring out intellectual property from your company, and the development and ownership of intellectual property in the future.

Step 2: Inventory the major <u>existing</u> product or service lines of business of your company.

Step 3: Correlate your intellectual property with the products and services that are protected by the intellectual property. From the opposite point of view, also correlate your lines of business with the intellectual property protecting each line of business from avoidable competition.

Step 4: Determine if any of your products or services are unprotected from avoidable competition by intellectual property. If so, develop an intellectual property strategy to protect each unprotected line of business. Be sure to apply this process to new lines of business that are not yet for sale but planned and under development.

Rule of Thumb 1

Each product should have at least one intellectual property protecting that product from competition. Utility patents, where they can be obtained, may be the best form of protection available.

Step 5: Plot the growth of your intellectual property, especially your patents, over time. As a benchmark, determine whether your rate of intellectual property portfolio growth is keeping up with that of your competitors.

Step 6: Determine the major lines of business of your competitors.

Step 7: Determine what intellectual property, if any, is protecting each of your competitor's lines of business.

Step 8: Develop a strategy to defeat any intellectual property protecting each of your competitor's lines of business from your competition. In the case of competitor patents, determine if they can be invented around.

Market Driven Intellectual Property Audit for <u>New</u> Products and Services

Step 1: List all your <u>new</u> lines of business (goods or services) currently under development.

Step 2: Itemize your intellectual property strategy for protecting each of these new lines of business from competition.

Step 3: Determine the risk for each new line of business of suppression by competitors asserting infringement of competitors' intellectual property. Where the risks are serious, determine a strategy to legitimately circumvent the competitor's intellectual property.

Technology Driven Intellectual Property Audit

Step 1: Inventory each research and development effort at your company.

Step 2: Specify the intellectual property for each project under development, both offensive (to protect from copycat competition by competitors), and defensive (to avoid infringement attacks by your competitors enforcing their intellectual property).

Step 3: For unprotected R&D projects, develop an adequate intellectual property strategy (offensive and defensive), or consider terminating the project. (Sell what you can patent, patent what you can sell.)

Rule of Thumb 2 (For Manufacturing).

A rule of thumb in some manufacturing industries is to develop one patent for each $1 million of research and development funds expended.

Licensing

Step 1: From your intellectual property inventory, determine which intellectual properties do not cover a current or planned line of business. Determine for each of these intellectual properties if you may sell or license that intellectual property to a non-competitor to generate cash flow.

Step 2: Regarding each intellectual property that covers a current or planned good or service, determine if that intellectual property may be licensed on a non-exclusive basis for cash to use in a non-competitive way.

Patent Cluster Analysis

Step 1: List your current patents.

Step 2: Check the appropriate data bases to determine which of your patents are referred to as named prior art references in other patents. Develop a "lineage" chart of patent cross references to prior art.

Step 3: Determine the owners of the various patents in the cross reference lineage chart.

Step 4: Determine if any of your patents are developing a cluster of patents around them from a particular competitor referring to your individual patent. This may indicate that a competitor is attempting a "picket fence" strategy to contain the utility of one of your key patents. (The "picket fence" strategy is discussed elsewhere in Chapter 3 of *Patent Strategies for Business, third edition*, the earlier companion volume of this book).

Step 5: If any of your key patents are being "picket fenced", determine if the competitor's picket fence patents can be "leap frogged". (The "leap frog strategy" is discussed elsewhere in Chapter 3 of *Patent Strategies for Business, third edition*, the earlier companion volume of this book.)

Step 6: Determine if you can "picket fence" any of your competitor patents in your industry.

9

Tech Trends: Telecom Services

Large profit opportunities can be developed by connecting the technology trends in this chapter (and subsequent chapters of this book), with the rules for inventing and patenting in Chapters 2 and 3 of *Patent Strategies for Business, third edition* (the companion volume to this book), and the rules for inventing and patenting for software and the Internet in Chapter 5 of this book.

Technology

The world is shifting from private switched telephone networks ("PSTN") and other circuit switched networks for telecom services to data packet based networks using ATM (asynchronous transfer mode) switches. This trend will continue and be driven by growth in Internet use, TCP/IP protocol use, and data traffic. Legacy telecom providers will continue to convert their old circuit switched networks to data packet digital networks. This will facilitate the further growth of VPN (virtual private network) applications and e-commerce of all types.

As voice over IP systems proliferate, additional intelligence will be provided to voice communication, including voice to text, text to voice, and hands-free voice interfaces.

There will be an increase in competition and urgency in the search for the profitable business model to provide broad digital bandwidth to as many users as possible. The big mystery at this point is how broad digital bandwidth will be first successfully provided to a broad base of residential users at a reasonable price. The players who bet on the right model will lock into market share and make fortunes. In any case, the cost of digital access will continue to go down as bandwidth goes up, and the telecom services industry will continue consolidating.

Services

The consolidation of the telecom services industry and technological changes will increasingly allow all telecom services to be provided by a smaller number of providers in a single billing format. Voice, video, fax, data and the Internet will eventually all be transmitted in a digital packet system by a single billing provider at a reduced cost.

Local exchange carriers will continue to dominate the local telephone access system. However, competition for access to residential users will, to some extent, come from alternative service providers through the local telephone access system. Furthermore, competition to access residential users increasingly will come from alternative network technologies, such as wireless ground transmission, satellite transmission, cable telephony, and power line transmission.

New telecom services will be enabled and provided by a dynamic evolving telecom infrastructure, including user-based (roaming, broadly construed, that is, the "number" follows the user) telecom services. (In effect, this has happened through the Internet with e-mail.)

New pricing models will be presented for telecom services, for example based on flat rates, time, usage, system traffic, or bandwidth, and will accommodate multiple qualities of service. The old telecom service pricing based on distance and time of day will continue to fade away.

Competitive Deregulation – and the Winners

Throughout the world, the trend will continue to privatize and deregulate telecom service providers. Where legacy systems are protected, investment and development of telecom systems will lag behind those in other more deregulated environments. Service, investment funds, and new technology will seek those jurisdictions that provide the most beneficial deregulatory environment.

The legal and political parameters of deregulation will dictate the speed of change and development in the telecom industry in each jurisdiction, and will profoundly influence the profitable business models that dominate in each jurisdiction.

Technical innovation, investment, and profits will be most often found in unregulated private environments that are free of government intervention seeking government tax revenue or social engineering goals (such as universal service).

In the first years of the new millennium, telecom services will be dominated around the world by from four to seven global carriers or alliances. (Major candidates include AT&T, MCI WorldCom, Sprint, British Telecom, and Telefonica S.A. Bell Atlantic and SBC may make the cut.) In addition, there will be a wide variety of niche players.

Much of the new technological development and development of new telecom services will be done by independent start-up companies, and companies new to the telecom service industry that are established in other industries (especially software and computers). These innovative niche players will eventually become allied with or merge into larger global super players. The larger telecom service legacy players will not be dominant innovators in many new telecom services, at least in the early days before these new markets are proven. Eventually these large incumbent players will use their huge financial strength to buy into proven new markets and services.

10

Tech Trends: the Internet, Intranets and Extranets

Large profit opportunities can be developed by connecting the technology trends in this chapter with the rules for inventing and patenting in Chapter 2 and 3 of *Patent Strategies for Business, third edition*, (the earlier companion volume of this book), and the rules for inventing and patenting for software and the Internet in Chapter 5 of this book.

Growth and Protocols

The popularity of the Internet and practical applications for the technology will continue to increase rapidly. New goods and services will be enabled and delivered by Internet technology.

Growth of the Internet will include emerging countries and will not be limited to the U.S. and Europe.

The Internet will be increasingly trusted for confidential and secure transactions, including e-commerce, using software technology such as SSL (Secure Socket Layer), SET (Secure Electronic Transaction), digital certificates, and other practical applications of encryption. Secure access over the Internet to confidential databases of great value for electronic commerce, medicine, research, and otherwise, will become accepted. This will greatly expand the use of extranets to allow customers and vendors to enter corporate intranets (to access data bases and execute transactions) on a limited authorization basis.

The Internet will not become a real time entertainment medium to compete directly with television or movies in the very immediate future. This will not happen until common use develops of new higher level Internet protocols that provide a guaranteed priority level of service for a real time critical function.

Streaming multimedia protocols will be used for distribution of prerecorded multimedia programming, and for video conferencing and telecommunication within corporate users that will accept a lower quality of service. Eventual distribution of protocols such as Internet Protocol Version 6 may provide the quality of service guarantees for real time streaming multimedia formats that are necessary to establish wide acceptance of this environment.

The public Internet will use improved Internet protocols including guaranteed priority, improved quality of service for real time formats, and bandwidth reservation. However, in the immediate future the public Internet still will not be accepted for real time critical or mission critical functions. For such critical functions, private wide area networks will continue to be used.

Businesses will make increased use of fax over Internet and voice over Internet. Fax over Internet will be a bigger use than voice over Internet in the short term, because fax over Internet is less real time critical. Some corporations and consumer hobbyists will accept lower immediate quality of service for voice over Internet.

Backbone and Service

Cable modems, DSL (digital subscriber lines), and other high speed technologies in the local loop will enable new Internet based multimedia and entertainment services, in areas where such infrastructure is available. Most consumer users of the Internet will continue to have access through conventional slow analog/dial-up modems through twisted copper pairs, which will limit the new Internet broad band services that are practically available.

As the Internet becomes more critical to business functions, steps will be taken to increase the physical security of the Internet infrastructure from attack or malfunction.

New Protocols and Standards

A rapid evolution of Internet standard protocols will continue, being pushed for demand for improved services. This will improve development environments, browsers, and directory services. Standardization will be pushed by informal bodies such as IETF (Internet Engineering Task Force) and W3C (Worldwide Web Consortium) rather than international standards bodies.

Web browsers will use digital certificates, such as the X.509 standard digital certificate, to authenticate users, in addition to standard user names and passwords.

More web pages will use DHTML (Dynamic HTML) and CSS (Cascading Style Sheet) extensions to HTML 4.0, in addition to XML (Extensible Markup Language). These software technologies will increase the flexibility of web pages.

XML will facilitate business applications communicating with each other over the Internet by allowing exchange of structured information on an application to application basis. This will allow access to services and information through public web pages without reverse engineering or recreating the website's data formats.

Web Servers

HTML (Hyper Text Markup Language) will become a basic function of all server platforms and additional services will be built on the HTTP (Hyper Text Transfer Protocol) protocol. These servers will improve the support of customized personal websites.

Major operating systems will include web server capabilities, improving scalability and reliability, and enabling large transaction volumes at the enterprise level.

More servers will be used that are HTTP/1.1 compliant to facilitate proxy caching and persistent connections. This will allow caching of content received over the Internet inside the fire wall, reducing overhead and resource utilization of network assets.

A New Profit Model

A startling development that will increase and be pushed by the Internet will be the new Internet software profit model. This calls for the free distribution of client software, or at least the low cost distribution of the same for business users. Profits in this "giveaway" software model, if any, will come from provision of services through web links built into the free software, including advertising and selling of server applications supporting the free client applications. In effect, giving away the client applications will promote market share and sales for the server applications.

Web Browsers

A web browser interface will become the preferred interface of all network applications, including corporate and data base applications. This will dictate which future hardware and software platforms are purchased in the marketplace, and diminish the apparent differences between client side operating systems.

The web browser market will continue to be dominated by Netscape and Microsoft. Microsoft will continue to increase its market share in this market at a rate affected by the outcome of the pending U.S. antitrust action against Microsoft. In any case, Microsoft will continue to offer browsers. Optimization of compatibility of applications to the Internet Explorer browser by Microsoft will be more important than strict compatibility with other third party industry protocols and standards.

The government's role in developing new Internet standards will continue to erode.

Client side applications will continue to incorporate browser functions. The degree of integration of client applications with browser features will be affected by the outcome of the current Microsoft antitrust action.

Browsers will be further fortified to facilitate navigation of the Web. Techniques will include client search engines, intelligent agents, Web site map outlining and visualization, and other features. Browsers will increase their support of push technology over the Internet.

Internet appliances (smart equipment with less intelligence than PCs) with selected browser functions will become more common. These will include mobile wireless devices, and wired devices. They may include PDAs (personal digital assistants), palm top computers, telephones, pagers, televisions, VCRs, home security systems, video cameras, HVAC (Heating Ventilation Air Conditioning) systems, lighting, and others.

11

Tech Trends: E-Commerce

Large profit opportunities can be developed by connecting the technology trends in this chapter with the rules for inventing and patenting in Chapters 2 and 3 of *Patent Strategies for Business, third edition*, (the companion volume of this book), and the rules for inventing and patenting for software and the Internet in Chapter 5 of this book.

Electronic commerce using the Internet will continue to grow rapidly. There will be more Internet users, improved security, and a larger number of off the shelf e-commerce software applications. All of these factors will drive increased Internet e-commerce.

E-commerce developments will become a fundamental strategic issue for many businesses. Note that in *The Wall Street Journal*, May 10, 1999, page A4, the President of Bank One indicated that he had no further interest in expansion through acquisitions. Instead, he would expand the bank in the future by Internet e-commerce strategies. He mentioned that this would enable a global expansion of his business without bank acquisitions, and that patent strategies for new lines of business and methods for delivering business services will continue to be critical to the market share of Internet e-commerce developments. Also, on the cover of *Forbes* magazine for May 17, 1999, an article was featured entitled "New Age Edison" regarding Jay Walker and Priceline.com. The sub-heading on the cover indicates that Priceline.com has a strategy "to patent Internet business models". Furthermore, in *The Wall Street Journal* of May 7, 1999, page B1, there was an article entitled "Two Entrepreneurs Try to Turn Net Patent Into a Blockbuster", indicating how a company is seeking to dominate the industry of distribution of video, music, and other audio over the Internet through an Internet patent strategy.

E-commerce revenue will continue to grow quickly, with transactions between businesses being of a greater growth, value, and revenue than transactions with consumers. The per transaction value of transactions among businesses will be greater than of transactions with consumers.

An increasing number of businesses of all sizes will be involved in e-commerce. This will be facilitated by more options to contract out Internet related services, including transaction management, security, installation and maintenance of systems and networks, and related hosting services.

New business models of e-commerce will be developed using features unique to the Internet. Old business models will be adapted to the Internet, if they can offer greater choice from a more extensive "inventory", greater convenience, and in some cases better prices to consumers.

One successful business model that will continue to grow will be the subsidy by advertising revenue to "free" web services (particularly for the distribution of information).

E-commerce Between Businesses

An increasing number of large corporations will require e-commerce interfaces for vendors to sell products to them. This will encourage the further growth of e-commerce transactions between businesses. Basically, large corporations will force their vendors to use a buyer promoted Web-based procurement system.

E-commerce purchasing of raw materials and other inputs used by a company will grow; however, this growth will be less than the growth of e-commerce buying of MRO (maintenance, repair and operations) materials.

An increasing number of ERP (enterprise resource planning) software systems will incorporate web enabled procurement functions.

Most transactions in e-commerce between businesses will be through access to on-line Internet or extranet catalogs. Direct communication between in-house software applications such as inventory, purchasing, and accounting will lag the development of the on-line catalog interactions.

An increasing number of complex products will be configured and ordered through e-commerce, without human assistance, by increasingly sophisticated programmed product configuration and ordering systems. One example of this in the retail field may be Dell.com's computer ordering functions. This approach will tie into the "zero inventory" that allows Dell to use real time purchase order completion as an alternative to inventory prediction and maintenance.

Internet communications will continue to grow and dominate the e-commerce field, replacing and merging with the pre-exiting EDI (electronic data interchange) techniques and providers. (See EDI service providers such as GEIS, IBM Global Services, and Sterling Commerce.) New users to e-commerce between businesses, especially smaller suppliers, will begin by using Internet applications rather than VAN (value added networks) provided by the original EDI service providers.

Specific industries will develop their own industry standards for the organization and transformation of data and data formats in Internet and web based electronic commerce. XML (extensible markup language) will dominate the evolution of e-commerce and the migration of EDI to the web.

E-commerce with Consumers

Web sites oriented towards consumer transactions will continue to evolve to be more user friendly. This will include Web design intended for faster download, and individually distinguished Web page look and feel for each user. Continuing efforts will be made to promote brand loyalty towards individual consumer Web sites.

Consumer oriented e-commerce will continue to be constrained by a lack of high speed Internet access. This will be caused both by congestion on the network backbone, and limited bandwidth close to the end consumer, in the local loop.

Consumers will continue to be concerned with confidentiality of use of their personal information delivered over the Web, most particularly their credit card numbers. However, increasingly users will become more confident of the confidentiality of the system, and in particular will come to view the use of credit cards and e-commerce over the Internet as at least as secure as normal credit card use with people at retail establishments.

Payment Systems for E-commerce

For e-commerce between businesses, corporate credit cards to authorized buyers will be used for an increasing number of high volume, low value purchases. Additional software systems may be developed to facilitate these functions. Many business to business e-commerce systems will use existing payment processes.

Payment processes will continue to evolve out of the BIPS (Bank Internet Payment System).

The SET protocol will expand to include smart cards and debit cards. Additional SET (Secure Electronic Transaction) merchants and SET banks will continue to appear, but at a rate behind the rate in Europe. The SET standard will be promoted and expanded, but may not dominate the market in the near term.

Most consumer oriented e-commerce transactions will use credit and debit cards in the traditional payment system. Payment systems will be developed to use e-money and smart cards (stored value card), as systems are developed and rolled out.

Most credit card payments with consumers will continue to use the SSL (secure socket layer) encrypted transmission of credit card information. Software systems will be added for fraud detection and blockage, including sophisticated artificial intelligent detection systems (such as using neural networks), in addition to less intelligent systems (for example, using address verification, mother's maiden name verification, passwords, or otherwise).

Software for E-Commerce

Major ERP (Enterprise Resource Planning) software vendors will incorporate e-commerce functions into their suites of products. Initially these will focus on procurement functions. Consolidation in this industry will continue with the acquisition of new innovative niche players by the established market leaders.

Industry wide (public domain) standards will fail to dominate e-commerce software, and proprietary vendor GUIs (Graphical User Interface) will dominate the market. One exception to this general trend may be the incorporation of third party software components for payment processing. Proprietary standards by specific e-commerce software vendors will continue to dominate the market over industry standards, and the early capture of market share by proprietary standards will continue to suppress introduction of industry standards.

It will be a continuing problem for corporations to integrate e-commerce functions into their software applications. This effort will be facilitated by corporations that use application suites from major vendors that incorporate their own vendor e-commerce interfaces.

Market Consolidation of Distribution Chains and Reorganization of Business

Small start-ups will provide much of the innovation in many e-commerce niches, and they will capture market share for their e-commerce applications.

Financial services, telecom services, insurance products, and retail goods will be marketed directly to consumers. This will allow the sources of the goods and services to compress and eliminate layers in the distribution chain (and the resulting costs), to provide their goods and services directly to retail users, at a reduced cost to the consumer and an increased profit margin to the provider. Many middle men in these industries will be put out of business.

These growing e-services will include traditional services such as life insurance, small business loans, mortgages, and banking. E-services will also include new services enabled only by e-commerce strategies, such as electronic bill payment and presentment.

Furthermore, the e-service strategy will allow non-traditional players in financial services to enter the financial service field. For example, "non-banks" will provide an increasing number of banking services. The hyperbole "banking is essential, but banks are not" will look less exaggerated as non-banks expand their activities. Traditional service providers, particularly in banking and financial services, will attack this new competitive threat by acquiring new innovative niche e-service providers.

Regulation of E-Commerce

Currently, e-commerce avoids much taxation, partly from jurisdictional and policing problems, and partly from government disinterest. However, as the financial volume of e-commerce explodes, it will inevitably attract increasing government interest to tax this potential source of revenue. No government can for long ignore a new target of taxation. (Governments seek new "markets" for their taxes, and more "market share" of the economy.) Because e-commerce is inherently international in scope, questions of tax harmonization, VAT taxes, customs, and tariffs will become issues at the World Trade Organization, the Organization of Economic Cooperation and Development, the European Union, and other international bodies.

The efforts of government to control e-commerce will lag behind the explosive growth of the industry for the near term.

12

Tech Trends: Smart Cards

The rest of the world outside the U.S. will continue to apply smart card technology more quickly than in the U.S. This will be promoted outside the U.S. by active central banking authorities and by the relative lack of an available and a cheap telecom system of real time debit and credit card transactions.

Interoperability between smart card technologies and protocols will continue to be a barrier to universal smart card acceptance. However, use of smart cards in closed institutional or special purpose environments will expand in the U.S.

Various countries outside the U.S. will introduce universal smart card systems for multiple functions, including financial, government, and medical functions.

In the U.S. smart cards will be applied broadly for access to corporate computer networks.

The EMV protocol (Europay/Mastercard/Visa Consortium Protocol) will become a major contender for an industry standard for smart card applications to debt and credit card transactions.

Object oriented software environments similar to Java will become increasingly common in smart card environments.

Run-time environments that download software applications on an as-needed basis, for example JavaCard, will eventually be applied to a greater extent in smart card applications, particularly those involving multiple use cards.

JavaCard and MultOs will become the major contenders for an industry standard for smart card operating systems. A common API (Application Program Interface) set may be developed that will run on both such systems, further facilitating the applications of smart cards and their acceptance.

Wireless smart card transactions will be promoted by exploitation of existing card functionality in mobile telephone handsets, in the SIM (Subscriber Identity Module). This will allow GSM (Global System for Mobile Communication) telephone handsets to be used as general purpose wireless smart card devices.

Customer loyalty programs will be combined with e-money functions in smart cards issued by private companies, such as financial institutions and retailers, to their customers.

Smart cards that can be read at a distance ("contactless" smart cards) will be used to track almost anything that moves (for example, railroad cars, trucks, or warehouse pallets) in various logistical and toll collection applications.

13

Tech Trends:
New Devices and Software

GPS

GPS (Global Positioning System) units will continue to become smaller, cheaper and require less power. These will find increasing uses integrated with a variety of mobile equipment, including mobile phones, PDA's (Personal Digital Assistants), other hand held telecom units, motor vehicles, and mobile equipment. This will permit the development of new collateral telecom services and equipment functions.

Current GPS for civilians has an accuracy of only 70 to 100 meters. Widespread availability of differential GPS will increase this accuracy to millimeters. GPS enabled chip sets will be available at a cost of about $10 per unit, and will be embedded in a wide variety of equipment. GPS combined with digital mapping and wireless telecom will result in locational readouts in user friendly map and address format, with directions, instead of merely in difficult GPS longitude and latitude format. The FCC (Federal Communications Commission) has mandated automatic GPS data with 911 emergency cell phone calls by 2001.

Killer App

Video-on-demand is <u>not</u> the Internet Killer Application. The Killer App is networking computers by VPN (virtual private networks), and by open Internet, including full duplex voice over IP (Internet telephony).

Video Cams

Video cameras and microphones developing a digital signal that can be modified and broadcast in the TCP/IP protocol will become increasingly small, cheap, and require less power. These will be integrated into a variety of existing systems facilitating new remote services.

Biometrics

Biometric devices and software will become increasingly effective and cheap. These will include iris scanning, fingerprints, voiceprints, hand geometry, face geometry, and signature authentication. These devices and techniques will be integrated into a variety of systems, offering increased levels of convenience and security.

Wireless Telecom

Wireless telecom service providers and equipment manufacturers will develop a single worldwide technical standard for mobile telecommunication devices. This will allow hand held devices to roam around the world, with high speed internet access, and to receive multimedia information and services. This will affect third generation (3G) wireless telecom systems, that is, the UMTS (Universal Mobile Telephone Service).

The leader in this development will be the Operators Harmonization Group, which currently includes AT&T Wireless, British

Telecom, Ericson, and Nortel Networks. This group will develop a common Internet backbone with the new standards using an TCP/IP packet switch system. The new protocol standards will facilitate a seamless transition from the variety of currently competing mobile telecom standards to a single world wide high speed standard. This will save billions of dollars for the telecom industry and expedite the distribution of its services.

Satellite Wireless

Broadband satellite wireless telecom ("Internet in the sky") will become common. This will be based on low earth orbit (LEO), non-geostationary (NGSO) satellite systems. Broadband LEO's will provide the wireless satellite equivalent of a fiber optic access link. LEO satellites are 96% closer to the earth than geostationary satellites (the latter have an altitude of 850 miles); therefore, LEO's eliminate latency and increase bandwidth in satellite wireless telecom.

Appendix 1

State Street Bank v. Signature Financial
149 F.3d 1368
47 U.S.P.Q.2d 1596

State Street Bank & Trust Co., Plaintiff-Appellee,
v.
Signature Financial Group, Inc. Defendant-Appellant.

No. 96-1327.

United States Court of Appeals,
Federal Circuit.

July 23, 1998.

*1369 Before RICH, PLAGER, and BRYSON, Circuit Judges.

RICH, Circuit Judge.

Signature Financial Group, Inc. (Signature) appeals from the decision of the United States District Court for the District of Massachusetts granting a motion for summary judgment in favor of State Street Bank & Trust Co. (State Street), finding U.S. Patent No. 5,193,056 (the '056 patent) invalid on the ground that the claimed subject matter is not encompassed by 35 U.S.C. § 101 (1994). See State Street Bank & Trust Co. v. Signature Financial Group, Inc., 927 F.Supp. 502, 38 USPQ2d 1530 (D.Mass.1996). We reverse and remand because we conclude that the patent claims are directed to statutory subject matter.

BACKGROUND

Signature is the assignee of the '056 patent which is entitled "Data Processing System for Hub and Spoke Financial Services Configuration." The '056 patent issued to Signature on 9 March 1993, naming R. Todd Boes as the inventor. The '056 patent is generally directed to a data processing system (the system) for implementing an investment structure which was developed for use in Signature's business as an administrator and accounting agent for mutual funds. In essence, the system, identified by the proprietary name Hub and Spoke (R), facilitates a structure whereby mutual funds (Spokes) pool their assets in an investment portfolio (Hub) organized as a partnership. This investment configuration provides the administrator of a mutual fund with the advantageous combination of economies of scale in administering investments coupled with the tax advantages of a partnership.

State Street and Signature are both in the business of acting as custodians and accounting agents for multi-tiered partnership fund financial services. State Street negotiated with Signature for a license to use its patented data processing system described and claimed in the '056 patent. When negotiations broke down, State Street brought a declaratory judgment action asserting invalidity, unenforceability, and noninfringement in Massachusetts district court, and then filed a motion for partial summary judgment of patent invalidity for failure to claim statutory subject matter under § 101. The motion was granted and this appeal followed.

DISCUSSION

On appeal, we are not bound to give deference to the district court's grant of summary judgment, but must make an independent determination that the standards for summary judgment have been met. Vas-Cath, Inc. v. Mahurkar, 935 F.2d 1555, 1560, 19 USPQ2d 1111, 1114 (Fed.Cir.1991). Summary judgment is properly granted where there are no genuine issues of material fact and the moving party is entitled to judgment as a matter of law. Fed.R.Civ.P. 56(c). The substantive issue at hand, whether the '056 patent is invalid for failure to claim statutory subject matter under § 101, is a matter of both claim construction and statutory construction. "[W]e review claim construction de novo including any allegedly fact-based questions relating to claim construction." Cybor Corp. v. FAS Techs., 138 F.3d 1448, 1451, 46 USPQ2d 1169, 1174 (Fed.Cir.1998) (in banc). We also

e-Patent Strategies

review statutory construction de novo. See Romero v. United States, 38 F.3d 1204, 1207 (Fed.Cir.1994). We hold that declaratory judgment plaintiff State Street was not entitled to the grant of summary judgment of invalidity of the '056 patent under § 101 as a matter of law, because the patent claims are directed to statutory subject matter.

The following facts pertinent to the statutory subject matter issue are either undisputed or represent the version alleged by the nonmovant. See Anderson v. Liberty Lobby, *1371 Inc., 477 U.S. 242, 255, 106 S.Ct. 2505, 91 L.Ed.2d 202 (1986). The patented invention relates generally to a system that allows an administrator to monitor and record the financial information flow and make all calculations necessary for maintaining a partner fund financial services configuration. As previously mentioned, a partner fund financial services configuration essentially allows several mutual funds, or "Spokes," to pool their investment funds into a single portfolio, or "Hub," allowing for consolidation of, inter alia, the costs of administering the fund combined with the tax advantages of a partnership. In particular, this system provides means for a daily allocation of assets for two or more Spokes that are invested in the same Hub. The system determines the percentage share that each Spoke maintains in the Hub, while taking into consideration daily changes both in the value of the Hub's investment securities and in the concomitant amount of each Spoke's assets.

In determining daily changes, the system also allows for the allocation among the Spokes of the Hub's daily income, expenses, and net realized and unrealized gain or loss, calculating each day's total investments based on the concept of a book capital account. This enables the determination of a true asset value of each Spoke and accurate calculation of allocation ratios between or among the Spokes. The system additionally tracks all the relevant data determined on a daily basis for the Hub and each Spoke, so that aggregate year end income, expenses, and capital gain or loss can be determined for accounting and for tax purposes for the Hub and, as a result, for each publicly traded Spoke.

It is essential that these calculations are quickly and accurately performed. In large part this is required because each Spoke sells shares to the public and the price of those shares is substantially based on the Spoke's percentage interest in the portfolio. In some instances, a mutual fund administrator is required to calculate the value of the shares to the nearest penny within as little as an hour and a half after the market closes. Given the complexity of the calculations, a computer or equivalent device is a virtual necessity to perform the task.

The '056 patent application was filed 11 March 1991. It initially contained six "machine" claims, which incorporated means-plus-function clauses, and six method claims. According to Signature, during prosecution the examiner contemplated a § 101 rejection for failure to claim statutory subject matter. However, upon cancellation of the six method claims, the examiner issued a notice of allowance for the remaining present six claims on appeal. Only claim 1 is an independent claim.

The district court began its analysis by construing the claims to be directed to a process, with each "means" clause merely representing a step in that process. However, "machine" claims having "means" clauses may only be reasonably viewed as process claims if there is no supporting structure in the written description that corresponds to the claimed "means" elements. See In re Alappat, 33 F.3d 1526, 1540-41, 31 USPQ2d 1545, 1554 (Fed.Cir.1994) (en banc). This is not the case now before us.

When independent claim 1 is properly construed in accordance with § 112, ¶ 6, it is directed to a machine, as demonstrated below, where representative claim 1 is set forth, the subject matter in brackets

e-Patent Strategies

stating the structure the written description discloses as corresponding to the respective "means" recited in the claims.

1. A data processing system for managing a financial services configuration of a portfolio established as a partnership, each partner being one of a plurality of funds, comprising:

(a) computer processor means [a personal computer including a CPU] for processing data;

(b) storage means [a data disk] for storing data on a storage medium;

(c) first means [an arithmetic logic circuit configured to prepare the data disk to magnetically store selected data] for initializing the storage medium;

(d) second means [an arithmetic logic circuit configured to retrieve information from a specific file, calculate incremental increases or decreases based on specific input, allocate the results on a percentage basis, and store the output in a *1372 separate file] for processing data regarding assets in the portfolio and each of the funds from a previous day and data regarding increases or decreases in each of the funds, [sic, funds'] assets and for allocating the percentage share that each fund holds in the portfolio;

(e) third means [an arithmetic logic circuit configured to retrieve information from a specific file, calculate incremental increases and decreases based on specific input, allocate the results on a percentage basis and store the output in a separate file] for processing data regarding daily incremental income, expenses, and net realized gain or loss for the portfolio and for allocating such data among each fund;

(f) fourth means [an arithmetic logic circuit configured to retrieve information from a specific file, calculate incremental increases and decreases based on specific input, allocate the results on a percentage basis and store the output in a separate file] for processing data regarding daily net unrealized gain or loss for the portfolio and for allocating such data among each fund; and

(g) fifth means [an arithmetic logic circuit configured to retrieve information from specific files, calculate that information on an aggregate basis and store the output in a separate file] for processing data regarding aggregate year-end income, expenses, and capital gain or loss for the portfolio and each of the funds.

Each claim component, recited as a "means" plus its function, is to be read, of course, pursuant to § 112, ¶ 6, as inclusive of the "equivalents" of the structures disclosed in the written description portion of the specification. Thus, claim 1, properly construed, claims a machine, namely, a data processing system for managing a financial services configuration of a portfolio established as a partnership, which machine is made up of, at the very least, the specific structures disclosed in the written description and corresponding to the means-plus-function elements (a)-(g) recited in the claim. A "machine" is proper statutory subject matter under § 101. We note that, for the purposes of a § 101 analysis, it is of little relevance whether claim 1 is directed to a "machine" or a "process," as long as it falls within at least one of the four enumerated categories of patentable subject matter, "machine" and "process" being such categories.

This does not end our analysis, however, because the court concluded that the claimed subject matter fell into one of two alternative judicially- created exceptions to statutory subject matter. [FN1] The court refers to the first exception as the "mathematical algorithm" exception and the second exception as the "business method" exception. Section 101 reads:

> FN1. Indeed, although we do not make this determination here, the judicially created exceptions, i.e., abstract ideas, laws of nature, etc., should be applicable to all categories of statutory subject matter, as our own precedent suggests. See Alappat, 33 F.3d at 1542, 31 USPQ2d at 1556; see also In re Johnston, 502 F.2d 765, 183 USPQ 172 (CCPA 1974) (Rich, J., dissenting).

Whoever invents or discovers any new and useful process, machine, manufacture, or composition of matter, or any new and useful improvement thereof, may obtain a patent therefor, subject to the conditions and requirements of this title.

The plain and unambiguous meaning of § 101 is that any invention falling within one of the four stated categories of statutory subject matter may be patented, provided it meets the other requirements for patentability set forth in Title 35, i.e., those found in §§ 102, 103, and 112, ¶ 2. [FN2]

> FN2. As explained in In re Bergy, 596 F.2d 952, 960, 201 USPQ 352, 360 (CCPA 1979) (emphases and footnote omitted):
>
> The first door which must be opened on the difficult path to patentability is § 101 .... The person approaching that door is an inventor, whether his invention is patentable or not .... Being an inventor or having an invention, however, is no guarantee of opening even the first door. What kind of an invention or discovery is it? In dealing with the question of kind, as distinguished from the qualitative conditions which make the invention patentable, § 101 is broad and general; its language is: "any * * * process, machine, manufacture, or

composition of matter, or any * * * improvement thereof." Section 100(b) further expands "process" to include "art or method, and * * * a new use of a known process, machine, manufacture, composition of matter, or material." If the invention, as the inventor defines it in his claims (pursuant to § 112, second paragraph), falls into any one of the named categories, he is allowed to pass through to the second door, which is § 102; "novelty and loss of right to patent" is the sign on it. Notwithstanding the words "new and useful" in § 101, the invention is not examined under that statute for novelty because that is not the statutory scheme of things or the long-established administrative practice.

*1373 The repetitive use of the expansive term "any" in § 101 shows Congress's intent not to place any restrictions on the subject matter for which a patent may be obtained beyond those specifically recited in § 101. Indeed, the Supreme Court has acknowledged that Congress intended § 101 to extend to "anything under the sun that is made by man." Diamond v. Chakrabarty, 447 U.S. 303, 309, 100 S.Ct. 2204, 65 L.Ed.2d 144 (1980); see also Diamond v. Diehr, 450 U.S. 175, 182, 101 S.Ct. 1048, 67 L.Ed.2d 155 (1981). [FN3] Thus, it is improper to read limitations into § 101 on the subject matter that may be patented where the legislative history indicates that Congress clearly did not intend such limitations. See Chakrabarty, 447 U.S. at 308, 100 S.Ct. 2204 ("We have also cautioned that courts 'should not read into the patent laws limitations and conditions which the legislature has not expressed.' " (citations omitted)).

> FN3. The Committee Reports accompanying the 1952 Act inform us that Congress intended statutory subject matter to "include anything under the sun that is made by man." S.Rep. No. 82-1979 at 5 (1952); H.R.Rep. No. 82-1923 at 6 (1952).

The "Mathematical Algorithm" Exception

The Supreme Court has identified three categories of subject matter that are unpatentable, namely "laws of nature, natural phenomena, and abstract ideas." Diehr, 450 U.S. at 185, 101 S.Ct. 1048. Of particular relevance to this case, the Court has held that mathematical algorithms are not patentable subject matter to the extent that they are merely abstract ideas. See Diehr, 450 U.S. 175, 101 S.Ct. 1048, passim; Parker v. Flook, 437 U.S. 584, 98 S.Ct. 2522, 57 L.Ed.2d 451 (1978); Gottschalk v. Benson, 409 U.S. 63, 93 S.Ct. 253, 34 L.Ed.2d 273 (1972). In Diehr, the Court explained that certain types of mathematical subject matter, standing alone, represent nothing more than abstract ideas until reduced to some type of practical application, i.e., "a useful, concrete and tangible result." Alappat, 33 F.3d at 1544, 31 USPQ2d at 1557. [FN4]

> FN4. This has come to be known as the mathematical algorithm exception. This designation has led to some confusion, especially given the Freeman-Walter-Abele analysis. By keeping in mind that the mathematical algorithm is unpatentable only to the extent that it represents an abstract idea, this confusion may be ameliorated.

Unpatentable mathematical algorithms are identifiable by showing they are merely abstract ideas constituting disembodied concepts or truths that are not "useful." From a practical standpoint, this means that to be patentable an algorithm must be applied in a "useful" way. In Alappat, we held that data, transformed by a machine through a series of mathematical calculations to produce a smooth waveform display on a rasterizer monitor, constituted a practical application of an abstract idea (a mathematical algorithm, formula, or calculation), because it produced "a useful, concrete and tangible result"--the smooth waveform.

Similarly, in Arrhythmia Research Technology Inc. v. Corazonix Corp., 958 F.2d 1053, 22 USPQ2d 1033 (Fed.Cir.1992), we held that the transformation of electrocardiograph signals from a patient's heartbeat by a machine through a series of mathematical calculations constituted a practical application of an abstract idea (a mathematical algorithm, formula, or calculation), because it corresponded to a useful, concrete or tangible thing--the condition of a patient's heart.

Today, we hold that the transformation of data, representing discrete dollar amounts, by a machine through a series of mathematical calculations into a final share price, constitutes a practical application of a mathematical algorithm, formula, or calculation, because it produces "a useful, concrete and tangible result"--a final share price momentarily fixed for recording and reporting purposes and even accepted and relied upon by regulatory authorities and in subsequent trades.

The district court erred by applying the Freeman-Walter-Abele test to determine whether the claimed subject matter was an unpatentable abstract idea. The Freeman-Walter-Abele test was designed by the Court *1374 of Customs and Patent Appeals, and subsequently adopted by this court, to extract and identify unpatentable mathematical algorithms in the aftermath of Benson and Flook. See In re Freeman, 573 F.2d 1237, 197 USPQ 464 (CCPA 1978) as modified by In re Walter, 618 F.2d 758, 205 USPQ 397 (CCPA 1980). The test has been thus articulated:

> First, the claim is analyzed to determine whether a mathematical algorithm is directly or indirectly recited. Next, if a mathematical algorithm is found, the claim as a whole is further analyzed to determine whether the algorithm is "applied in any manner to physical elements or process steps," and, if it is, it "passes muster under § 101."

e-Patent Strategies

In re Pardo, 684 F.2d 912, 915, 214 USPQ 673, 675-76 (CCPA 1982) (citing In re Abele, 684 F.2d 902, 214 USPQ 682 (CCPA 1982)). [FN5]

FN5. The test has been the source of much confusion. In In re Abele, 684 F.2d 902, 214 USPQ 682 (CCPA 1982), the CCPA upheld claims applying "a mathematical formula within the context of a process which encompasses significantly more than the algorithm alone." Id. at 909. Thus, the CCPA apparently inserted an additional consideration--the significance of additions to the algorithm. The CCPA appeared to abandon the application of the test in In re Taner, 681 F.2d 787, 214 USPQ 678 (CCPA 1982), only to subsequently "clarify" that the Freeman-Walter-Abele test was simply not the exclusive test for detecting unpatentable subject matter. In re Meyer, 688 F.2d 789, 796, 215 USPQ 193, 199 (CCPA 1982).

After Diehr and Chakrabarty, the Freeman-Walter-Abele test has little, if any, applicability to determining the presence of statutory subject matter. As we pointed out in Alappat, 33 F.3d at 1543, 31 USPQ2d at 1557, application of the test could be misleading, because a process, machine, manufacture, or composition of matter employing a law of nature, natural phenomenon, or abstract idea is patentable subject matter even though a law of nature, natural phenomenon, or abstract idea would not, by itself, be entitled to such protection. [FN6] The test determines the presence of, for example, an algorithm. Under Benson, this may have been a sufficient indicium of nonstatutory subject matter. However, after Diehr and Alappat, the mere fact that a claimed invention involves inputting numbers, calculating numbers, outputting numbers, and storing numbers, in and of itself, would not render it nonstatutory subject matter, unless, of course, its operation does not produce a "useful, concrete and tangible result." Alappat, 33 F.3d at 1544, 31 USPQ2d at 1557. [FN7] After all, as we have repeatedly stated,

FN6. See e.g. Parker v. Flook, 437 U.S. 584, 590, 98 S.Ct. 2522, 57 L.Ed.2d 451 (1978) ("[A] process is not unpatentable simply because it contains a law of nature or a mathematical algorithm."); Funk Bros. Seed Co. v. Kalo Inoculant Co., 333 U.S. 127, 130, 68 S.Ct. 440, 92 L.Ed. 588 (1948) ("He who discovers a hitherto unknown phenomenon of nature has no claim to a monopoly of it which the law recognizes. If there is to be invention from such a discovery, it must come from the application of the law to a new and useful end."); Mackay Radio & Tel. Co. v. Radio Corp. of Am., 306 U.S. 86, 94, 59 S.Ct. 427, 83 L.Ed. 506 (1939) ("While a scientific truth, or the mathematical expression of it, is not a patentable invention, a novel and useful structure created with the aid of knowledge of scientific truth may be.").

[W]hen a claim containing a mathematical formula implements or applies that formula in a structure or process which, when considered as a whole, is performing a function which the patent laws were designed to protect (e.g., transforming or reducing an article to a different state or thing), then the claim satisfies the requirements of § 101. Diehr, 450 U.S. at 192, 101 S.Ct. 1048; see also In re Iwahashi, 888 F.2d 1370, 1375, 12 USPQ2d 1908, 1911 (Fed.Cir.1989); Taner, 681 F.2d at 789, 214 USPQ at 680. The dispositive inquiry is whether the claim as a whole is directed to statutory subject matter. It is irrelevant that a claim may contain, as part of the whole, subject matter which would not be patentable by itself. "A claim drawn to subject matter otherwise statutory does not become nonstatutory simply because it uses a mathematical formula, computer program or digital computer." Diehr, 450 U.S. at 187, 101 S.Ct. 1048.

FN7. As the Supreme Court expressly stated in Diehr, its own holdings in Benson and Flook "stand for no more than these long-established principles" that abstract ideas and natural phenomena are not patentable. Diehr, 450 U.S. at 185, 101 S.Ct. 1048 (citing Chakrabarty, 447 U.S. at 309, 100 S.Ct. 2204 and Funk Bros., 333 U.S. at 130, 68 S.Ct. 440.).

every step-by-step process, be it electronic or chemical or mechanical, involves an algorithm in the broad sense of the term. Since § 101 expressly includes processes as a category of inventions which may be patented and § 100(b) further defines the word "process" as meaning "process, art or *1375 method, and includes a new use of a known process, machine, manufacture, composition of matter, or material," it follows that it is no ground for holding a claim is directed to nonstatutory subject matter to say it includes or is directed to an algorithm. This is why the proscription against patenting has been limited to mathematical algorithms.... In re Iwahashi, 888 F.2d 1370, 1374, 12 USPQ2d 1908, 1911 (Fed.Cir.1989) (emphasis in the original). [FN8]

> FN8. In In re Pardo, 684 F.2d 912 (CCPA 1982), the CCPA narrowly limited "mathematical algorithm" to the execution of formulas with given data. In the same year, in In re Meyer, 688 F.2d 789, 215 USPQ 193 (CCPA 1982), the CCPA interpreted the same term to include any mental process that can be represented by a mathematical algorithm. This is also the position taken by the PTO in its Examination Guidelines, 61 Fed.Reg. 7478, 7483 (1996).

The question of whether a claim encompasses statutory subject matter should not focus on which of the four categories of subject matter a claim is directed to [FN9]--process, machine, manufacture, or composition of matter--but rather on the essential characteristics of the subject matter, in particular, its practical utility. Section 101 specifies that statutory subject matter must also satisfy the other "conditions and requirements" of Title 35, including novelty, nonobviousness, and adequacy of disclosure and notice. See In re Warmerdam, 33 F.3d 1354, 1359, 31 USPQ2d 1754, 1757-58 (Fed.Cir.1994). For purpose of our analysis, as noted above, claim 1 is directed to a machine programmed with the Hub and Spoke software and admittedly produces a "useful, concrete, and tangible result." Alappat, 33 F.3d at 1544, 31 USPQ2d at 1557. This renders it statutory subject matter, even if the useful result is expressed in numbers, such as price, profit, percentage, cost, or loss.

> FN9. Of course, the subject matter must fall into at least one category of statutory subject matter.

The Business Method Exception

As an alternative ground for invalidating the '056 patent under § 101, the court relied on the judicially-created, so-called "business method" exception to statutory subject matter. We take this opportunity to lay this ill-conceived exception to rest. Since its inception, the "business method" exception has merely represented the application of some general, but no longer applicable legal principle, perhaps arising out of the "requirement for invention"--which was eliminated by § 103. Since the 1952 Patent Act, business methods have been, and should have been, subject to the same legal requirements for patentability as applied to any other process or method. [FN10]

> FN10. As Judge Newman has previously stated,
>
> [The business method exception] is ... an unwarranted encumbrance to the definition of statutory subject matter in section 101, that [should] be discarded as error-prone, redundant, and obsolete. It merits retirement from the glossary of section 101.... All of the "doing business" cases could have been decided using the clearer concepts of Title 35. Patentability does not turn on whether the claimed method does "business" instead of something else, but on whether the method, viewed as a whole, meets the requirements of patentability as set forth in Sections 102, 103, and 112 of the Patent Act. In re Schra-

der, 22 F.3d 290, 298, 30 USPQ2d 1455, 1462 (Fed.Cir.1994) (Newman, J., dissenting).

The business method exception has never been invoked by this court, or the CCPA, to deem an invention unpatentable. [FN11] Application of this particular exception has always been preceded by a ruling based on some clearer concept of Title 35 or, more commonly, application of the abstract idea exception based on finding a mathematical algorithm. Illustrative is the CCPA's analysis in In re Howard, 55 C.C.P.A. 1121, 394 F.2d 869, 157 USPQ 615 (CCPA 1968), wherein the court affirmed the Board of Appeals' rejection of the claims for lack of novelty and found it unnecessary to reach the Board's section 101 ground that a method of doing business is "inherently unpatentable." Id. at 872, 55 C.C.P.A. 1121, 394 F.2d 869, 157 USPQ at 617. [FN12]

> FN11. See Rinaldo Del Gallo, III, Are "Methods of Doing Business" Finally out of Business as a Statutory Rejection?, 38 IDEA 403, 435 (1998).

> FN12. See also Dann v. Johnston, 425 U.S. 219, 96 S.Ct. 1393, 47 L.Ed.2d 692 (1976) (the Supreme Court declined to discuss the section 101 argument concerning the computerized financial record-keeping system, in view of the Court's holding of patent invalidity under section 103); In re Chatfield, 545 F.2d 152, 157, 191 USPQ 730, 735 (CCPA 1976); Ex parte Murray, 9 USPQ2d 1819, 1820 (Bd.Pat.App & Interf. 1988) ("[T]he claimed accounting method [requires] no more than the entering, sorting, debiting and totaling of expenditures as necessary preliminary steps to issuing an expense analysis statement ....") states grounds of obviousness or lack of novelty, not of non-statutory subject matter.

*1376 Similarly, In re Schrader, 22 F.3d 290, 30 USPQ2d 1455 (Fed.Cir.1994), while making reference to the business method exception, turned on the fact that the claims implicitly recited an abstract idea in the form of a mathematical algorithm and there was no "transformation or conversion of subject matter representative of or constituting physical activity or objects." 22 F.3d at 294, 30 USPQ2d at 1459 (emphasis omitted). [FN13]

> FN13. Any historical distinctions between a method of "doing" business and the means of carrying it out blur in the complexity of modern business systems. See Paine, Webber, Jackson & Curtis v. Merrill Lynch, 564 F.Supp. 1358, 218 USPQ 212 (D.Del.1983), (holding a computerized system of cash management was held to be statutory subject matter.)

State Street argues that we acknowledged the validity of the business method exception in Alappat when we discussed Maucorps and Meyer:

> Maucorps dealt with a business methodology for deciding how salesmen should best handle respective customers and Meyer involved a "system" for aiding a neurologist in diagnosing patients. Clearly, neither of the alleged "inventions" in those cases falls within any § 101 category.

Alappat, 33 F.3d at 1541, 31 USPQ2d at 1555. However, closer scrutiny of these cases reveals that the claimed inventions in both Maucorps and Meyer were rejected as abstract ideas under the mathematical algorithm exception, not the business method exception. See In re Maucorps, 609 F.2d 481, 484, 203 USPQ 812, 816 (CCPA 1979); In re Meyer, 688 F.2d 789, 796, 215 USPQ 193, 199 (CCPA 1982). [FN14]

> FN14. Moreover, these cases were subject to the Benson era Freeman-Walter-Abele test--in other words, analysis as it existed before Diehr and Alappat.

Even the case frequently cited as establishing the business method exception to statutory subject matter, Hotel Security Checking Co. v. Lorraine Co., 160 F. 467 (2d Cir.1908), did not rely on the exception to strike the patent. [FN15] In that case, the patent was found invalid for lack of novelty and "invention," not because it was improper subject matter for a patent. The court stated "the fundamental principle of the system is as old as the art of bookkeeping, i.e., charging the goods of the employer to the agent who takes them." Id. at 469. "If at the time of [the patent] application, there had been no system of bookkeeping of any kind in restaurants, we would be confronted with the question whether a new and useful system of cash registering and account checking is such an art as is patentable under the statute." Id. at 472.

> FN15. See also Loew's Drive-in Theatres v. Park-in Theatres, 174 F.2d 547, 552 (1st Cir.1949) (holding that the means for carrying out the system of transacting business lacked "an exercise of the faculty of invention"); In re Patton, 29 C.C.P.A. 982, 127 F.2d 324, 327-28 (CCPA 1942) (finding claims invalid as failing to define patentable subject matter over the references of record.); Berardini v. Tocci, 190 F. 329, 332 (C.C.S.D.N.Y.1911); In re Wait, 22 C.C.P.A. 822, 73 F.2d 982, 983 (CCPA 1934) ("[S]urely these are, and always have been, essential steps in all dealings of this nature, and even conceding, without holding, that some methods of doing business might present patentable novelty, we think such novelty is lacking here."); In re Howard, 55 C.C.P.A. 1121, 394 F.2d 869, 157 USPQ 615, 617 (CCPA 1968) ("[W]e therefore affirm the decision of the Board of Appeals on the ground that the claims do not define a novel process [so we find it] unnecessary to consider the issue of whether a method of doing business is inherently unpatentable."). Although a clearer statement was made in In re Patton, 29 C.C.P.A. 982, 127 F.2d 324, 327, 53 USPQ 376, 379 (CCPA 1942) that a system for transacting business, separate from the means for

carrying out the system, is not patentable subject matter, the jurisprudence does not require the creation of a distinct business class of unpatentable subject matter.

This case is no exception. The district court announced the precepts of the business method exception as set forth in several treatises, but noted as its primary reason for finding the patent invalid under the business method exception as follows:

> If Signature's invention were patentable, any financial institution desirous of implementing a multi-tiered funding complex modelled (sic) on a Hub and Spoke configuration would be required to seek Signature's permission before embarking on *1377 such a project. This is so because the '056 Patent is claimed [sic] sufficiently broadly to foreclose virtually any computer- implemented accounting method necessary to manage this type of financial structure.

927 F.Supp. 502, 516, 38 USPQ2d 1530, 1542 (emphasis added). Whether the patent's claims are too broad to be patentable is not to be judged under § 101, but rather under §§ 102, 103 and 112. Assuming the above statement to be correct, it has nothing to do with whether what is claimed is statutory subject matter.

In view of this background, it comes as no surprise that in the most recent edition of the Manual of Patent Examining Procedures (MPEP) (1996), a paragraph of § 706.03(a) was deleted. In past editions it read:

> Though seemingly within the category of process or method, a method of doing business can be rejected as not being within the statutory classes. See Hotel Security Checking Co. v. Lorraine Co., 160 F. 467 (2nd Cir.1908) and In re Wait, 24 USPQ 88, 22 C.C.P.A. 822, 73 F.2d 982 (1934).

e-Patent Strategies

MPEP § 706.03(a) (1994). This acknowledgment is buttressed by the U.S. Patent and Trademark 1996 Examination Guidelines for Computer Related Inventions which now read:

> Office personnel have had difficulty in properly treating claims directed to methods of doing business. Claims should not be categorized as methods of doing business. Instead such claims should be treated like any other process claims.

Examination Guidelines, 61 Fed.Reg. 7478, 7479 (1996). We agree that this is precisely the manner in which this type of claim should be treated. Whether the claims are directed to subject matter within § 101 should not turn on whether the claimed subject matter does "business" instead of something else.

CONCLUSION

The appealed decision is reversed and the case is remanded to the district court for further proceedings consistent with this opinion.

REVERSED and REMANDED.

Appendix 2

AT&T v. Excel
172 F.3d 1352
50 U.S.P.Q.2d 1447

AT&T CORP., Plaintiff-Appellant,
v.
EXCEL COMMUNICATIONS, INC., Excel Communications Marketing, Inc., and Excel Telecommunications, Inc., Defendants-Appellees.

No. 98-1338.

United States Court of Appeals, Federal Circuit.

April 14, 1999.

Before PLAGER, CLEVENGER, and RADER, Circuit Judges.

PLAGER, Circuit Judge.

This case asks us once again to examine the scope of section 1 of the Patent Act, 35 U.S.C. 101 (1994). The United States District Court for the District of Delaware granted summary judgment to *Excel Communications, Inc.*, *Excel Communications Marketing, Inc.*, and *Excel Telecommunications, Inc.* (collectively *"Excel"*), holding U.S. Patent No. 5,333,184 (the '184 patent) invalid under Section 101 for failure to claim statutory subject matter. See *AT&T Corp. v. Excel Communications, Inc.*, No. CIV.A.96-434-SLR, 1998 WL 175878, at *7 (D.Del. Mar.27, 1998). *AT&T Corp.* (*"AT&T"*), owner of the '184 patent, appeals. Because we find that the claimed subject matter is properly within the statutory scope of Section 101, we reverse the district court's judgment of invalidity on this ground and remand the case for further proceedings.

BACKGROUND

A.

The '184 patent, entitled "Call Message Recording for Telephone Systems," issued on July 26, 1994. It describes a message record for long-distance telephone calls that is enhanced by adding a primary interexchange carrier ("PIC") indicator. The addition of the indicator aids long-distance carriers in providing differential billing treatment for subscribers, depending upon whether a subscriber calls someone with the same or a different long-distance carrier.

The invention claimed in the '184 patent is designed to operate in a telecommunications system with multiple long-distance service providers. The system contains local exchange carriers ("LECs") and long-distance service (interexchange) carriers ("IXCs"). The LECs provide local telephone service and access to IXCs. Each customer has an LEC for local service and selects an IXC, such as *AT&T* or *Excel*, to be its primary long-distance service (interexchange) carrier or PIC. IXCs may own their own facilities, as does *AT&T*. Others, like *Excel*, called "resellers" or "resale carriers," contract with facility-owners to route their subscribers' calls through the facility- owners' switches and transmission lines. Some IXCs, including MCI and U.S. Sprint, have a mix of their own lines and leased lines.

*1354 The system thus involves a three-step process when a caller makes a direct-dialed (1+) long-distance telephone call: (1) after the call is transmitted over the LEC's network to a switch, and the LEC identifies the caller's PIC, the LEC automatically routes the call to the facilities used by the caller's PIC; (2) the PIC's facilities carry the call to the LEC serving the call recipient; and (3) the call recipient's LEC delivers the call over its local network to the recipient's telephone.

When a caller makes a direct-dialed long-distance telephone call, a switch (which may be a switch in the interexchange network) monitors and records data related to the call, generating an "automatic message account" ("AMA") message record. This contemporaneous message record contains fields of information such as the originating and terminating telephone numbers, and the length of time of the call. These message records are then transmitted from the switch to a message accumulation system for processing and billing.

Because the message records are stored in electronic format, they can be transmitted from one computer system to another and reformatted to ease processing of the information. Thus the carrier's AMA message subsequently is translated into the industry-standard "exchange message interface," forwarded to a rating system, and ultimately forwarded to a billing system in which the data resides until processed to generate, typically, "hard copy" bills which are mailed to subscribers.

B.

The invention of the '184 patent calls for the addition of a data field into a standard message record to indicate whether a call involves a particular PIC (the "PIC indicator"). This PIC indicator can exist in several forms, such as a code which identifies the call recipient's PIC, a flag which shows that the recipient's PIC is or is not a particular IXC, or a flag that identifies the recipient's and the caller's PICs as the same IXC. The PIC indicator therefore enables IXCs to provide differential billing for calls on the basis of the identified PIC.

The application that issued as the '184 patent was filed in 1992. The U.S. Patent and Trademark Office ("PTO") initially rejected, for reasons unrelated to Section 101, all forty-one of the originally filed claims. Following amendment, the claims were issued in 1994 in their present form. The '184 patent contains six independent claims, five method claims and one apparatus claim, and additional dependent claims. The PTO granted the '184 patent without questioning whether the claims were directed to statutory subject matter under Section 101.

AT&T in 1996 asserted ten of the method claims against *Excel* in this infringement suit. The independent claims at issue (claims 1, 12, 18, and 40) include the step of "generating a message record for an interexchange call between an originating subscriber and a terminating subscriber," and the step of adding a PIC indicator to the message record. Independent claim 1, for example, adds a PIC indicator whose value depends upon the call recipient's PIC:

> 1. A method for use in a telecommunications system in which interexchange calls initiated by each subscriber are automatically routed over the facilities of a particular one of a plurality of interexchange carriers associated with that subscriber, said method comprising the steps of:
>
>> <u>generating a message record for an interexchange call</u> between an originating subscriber and a terminating subscriber, and
>>
>> <u>including, in said message record, a primary interexchange carrier (PIC) indicator</u> having a value which is a <u>function of whether or not the interexchange carrier associated with said terminating subscriber is a predetermined one</u> of said interexchange carriers. (Emphasis added.)

Independent claims 12 and 40 add a PIC indicator that shows if a *1355 recipient's PIC is the same as the IXC over which that particular call is being made. Independent claim 18 adds a PIC indicator designed to show if the caller and the recipient subscribe to the same

e-Patent Strategies

IXC. The dependent claims at issue add the steps of accessing an IXC's subscriber database (claims 4, 13, and 19) and billing individual calls as a function of the value of the PIC indicator (claims 6, 15, and 21).

The district court concluded that the method claims of the '184 patent implicitly recite a mathematical algorithm. See *AT&T*, 1998 WL 175878, at * 6. The court was of the view that the only physical step in the claims involves data-gathering for the algorithm. See id. Though the court recognized that the claims require the use of switches and computers, it nevertheless concluded that use of such facilities to perform a non-substantive change in the data's format could not serve to convert non-patentable subject matter into patentable subject matter. See id. at *6-7. Thus the trial court, on summary judgment, held all of the method claims at issue invalid for failure to qualify as statutory subject matter. See id. at *7.

DISCUSSION

A.

Summary judgment is appropriate if there are no genuine issues of material fact and the moving party is entitled to judgment as a matter of law. See Fed.R.Civ.P. 56(c). We review without deference a trial court's grant of summary judgment, with all justifiable factual inferences drawn in favor of the party opposing the motion. See *Anderson v. Liberty Lobby, Inc.*, 477 U.S. 242, 255, 106 S.Ct. 2505, 91 L.Ed.2d 202 (1986).

The issue on appeal, whether the asserted claims of the '184 patent are invalid for failure to claim statutory subject matter under 35 U.S.C. 101, is a question of law which we review without deference. See *Arrhythmia Research Technology v. Corazonix Corp.*, 958 F.2d 1053, 1055-56, 22 USPQ2d 1033, 1035 (Fed.Cir.1992). In matters of statutory interpretation, it is this court's responsibility independently to determine what the law is. See *Hodges v. Secretary of the Dep't of Health & Human Servs.*, 9 F.3d 958, 960 (Fed.Cir. 1993).

B.

Our analysis of whether a claim is directed to statutory subject matter begins with the language of 35 U.S.C. 101, which reads:

> Whoever invents or discovers any new and useful process, machine, manufacture, or composition of matter, or any new and useful improvement thereof, may obtain a patent therefor, subject to the conditions and requirements of this title.

The Supreme Court has construed Section 101 broadly, noting that Congress intended statutory subject matter to "include anything under the sun that is made by man." See *Diamond v. Chakrabarty*, 447 U.S. 303, 309, 100 S.Ct. 2204, 65 L.Ed.2d 144 (1980) (quoting S.Rep. No. 82-1979, at 5 (1952); H.R.Rep. No. 82-1923, at 6 (1952)); see also *Diamond v. Diehr*, 450 U.S. 175, 182, 101 S.Ct. 1048, 67 L.Ed.2d 155 (1981). Despite this seemingly limitless expanse, the Court has specifically identified three categories of unpatentable subject matter: "laws of nature, natural phenomena, and abstract ideas." See *Diehr*, 450 U.S. at 185, 101 S.Ct. 1048.

In this case, the method claims at issue fall within the "process" [FN1] category of the four enumerated categories of patentable subject matter in Section 101. The district court held that the claims at issue, though otherwise within the terms of Section 101, implicitly recite a mathematical algorithm, see AT&T, 1998 WL 175878, at *6, and thus fall within the judicially created *1356 "mathematical algorithm" exception to statutory subject matter.

> FN1. "Process" is defined in 35 U.S.C. s 100(b) to encompass: "[a] process, art or method, and includes a new use of a known

process, machine, manufacture, composition of matter, or material."

A mathematical formula alone, sometimes referred to as a mathematical algorithm, viewed in the abstract, is considered unpatentable subject matter. See *Diamond v. Diehr*, 450 U.S. 175, 101 S.Ct. 1048, 67 L.Ed.2d 155 (1981); *Parker v. Flook*, 437 U.S. 584, 98 S.Ct. 2522, 57 L.Ed.2d 451 (1978); *Gottschalk v. Benson*, 409 U.S. 63, 93 S.Ct. 253, 34 L.Ed.2d 273 (1972). Courts have used the terms "mathematical algorithm," "mathematical formula," and "mathematical equation," to describe types of nonstatutory mathematical subject matter without explaining whether the terms are interchangeable or different. Even assuming the words connote the same concept, there is considerable question as to exactly what the concept encompasses. See, e.g., *Diehr*, 450 U.S. at 186 n. 9, 101 S.Ct. 1048 ("The term 'algorithm' is subject to a variety of definitions ... [Petitioner's] definition is significantly broader than the definition this Court employed in Benson and Flook."); accord *In re Schrader*, 22 F.3d 290, 293 n. 5, 30 USPQ2d 1455, 1457 n. 5 (Fed.Cir.1994).

This court recently pointed out that any step-by-step process, be it electronic, chemical, or mechanical, involves an "algorithm" in the broad sense of the term. See *State Street Bank & Trust Co. v. Signature Fin. Group, Inc.*, 149 F.3d 1368, 1374-75, 47 USPQ2d 1596, 1602 (Fed.Cir.1998), cert. denied, --- U.S. ----, 119 S.Ct. 851, 142 L.Ed.2d 704 (1999). Because Section 101 includes processes as a category of patentable subject matter, the judicially-defined proscription against patenting of a "mathematical algorithm," to the extent such a proscription still exists, is narrowly limited to mathematical algorithms in the abstract. See id.; see also *Benson*, 409 U.S. at 65, 93 S.Ct. 253 (describing a mathematical algorithm as a "procedure for solving a given type of mathematical problem").

Since the process of manipulation of numbers is a fundamental part of computer technology, we have had to reexamine the rules that govern the patentability of such technology. The sea-changes in both law and technology stand as a testament to the ability of law to adapt to new and innovative concepts, while remaining true to basic principles. In an earlier era, the PTO published guidelines essentially rejecting the notion that computer programs were patentable. [FN2] As the technology progressed, our predecessor court disagreed, and, overturning some of the earlier limiting principles regarding Section 101, announced more expansive principles formulated with computer technology in mind. [FN3] In our recent decision in *State Street*, this court discarded the so-called "business method" exception and reassessed the "mathematical algorithm" exception, see 149 F.3d at 1373-77, 47 USPQ2d at 1600-04, both judicially-created "exceptions" to the statutory categories of Section 101. As this brief review suggests, this court (and its predecessor) has struggled to make our understanding of the scope of Section 101 responsive to the needs of the modern world.

FN2. See, e.g., 33 Fed.Reg. 15581, 15609-10 (1968).

FN3. See *In re Tarczy-Hornoch*, 55 C.C.P.A. 1441, 397 F.2d 856, 158 USPQ 141 (CCPA 1968) (overruling the "function of a machine" doctrine); see also *In re Bernhart*, 57 C.C.P.A. 737, 417 F.2d 1395, 163 USPQ 611 (CCPA 1969) (discussing patentability of a programmed computer); *In re Musgrave*, 57 C.C.P.A. 1352, 431 F.2d 882, 167 USPQ 280 (CCPA 1970) (analyzing process claims encompassing computer programs). For a more detailed review of this history, with extensive citation to the secondary literature, see Justice Stevens's dissent in *Diehr*, 450 U.S. at 193, 101 S.Ct. 1048.

The Supreme Court has supported and enhanced this effort. In *Diehr*, the Court expressly limited its two earlier decisions in *Flook* and *Benson* by emphasizing that these cases did no more than confirm the "long-established principle" that laws of nature, natural phenomena, and abstract ideas are excluded from patent protection. 450 U.S.

at 185, 101 S.Ct. 1048. The *Diehr* *1357 Court explicitly distinguished *Diehr's* process by pointing out that "the respondents here do not seek to patent a mathematical formula. Instead, they seek patent protection for a process of curing synthetic rubber." Id. at 187, 101 S.Ct. 1048. The Court then explained that although the process used a well-known mathematical equation, the applicants did not "pre-empt the use of that equation." Id. Thus, even though a mathematical algorithm is not patentable in isolation, a process that applies an equation to a new and useful end "is at the very least not barred at the threshold by Section 101." Id. at 188, 101 S.Ct. 1048. In this regard, it is particularly worthy of note that the argument for the opposite result, that "the term 'algorithm' ... is synonymous with the term 'computer program,' " id. at 219, 101 S.Ct. 1048 (Stevens, J., dissenting), and thus computer- based programs as a general proposition should not be patentable, was made forcefully in dissent by Justice Stevens; his view, however, was rejected by the *Diehr* majority.

As previously noted, we most recently addressed the "mathematical algorithm" exception in *State Street*. See 149 F.3d at 1373-75, 47 USPQ2d at 1600- 02. In *State Street*, this court, following the Supreme Court's guidance in *Diehr*, concluded that "[u]npatentable mathematical algorithms are identifiable by showing they are merely abstract ideas constituting disembodied concepts or truths that are not 'useful.' ... [T]o be patentable an algorithm must be applied in a 'useful' way." Id. at 1373, 47 USPQ2d at 1601. In that case, the claimed data processing system for implementing a financial management structure satisfied the Section 101 inquiry because it constituted a "practical application of a mathematical algorithm, ... [by] produc[ing] 'a useful, concrete and tangible result.' " Id. at 1373, 47 USPQ2d at 1601.

The *State Street* formulation, that a mathematical algorithm may be an integral part of patentable subject matter such as a machine or process if the claimed invention as a whole is applied in a "useful" manner, follows the approach taken by this court en banc in *In re*

Alappat, 33 F.3d 1526, 31 USPQ2d 1545 (Fed.Cir.1994). In *Alappat*, we set out our understanding of the Supreme Court's limitations on the patentability of mathematical subject matter and concluded that:

> [The Court] never intended to create an overly broad, fourth category of [mathematical] subject matter excluded from Section 101. Rather, at the core of the Court's analysis ... lies an attempt by the Court to explain a rather straightforward concept, namely, that certain types of mathematical subject matter, <u>standing alone</u>, represent nothing more than <u>abstract ideas until reduced to some type of practical application</u>, and thus that subject matter is not, in and of itself, entitled to patent protection.

Id. at 1543, 31 USPQ2d at 1556-57 (emphasis added). Thus, the *Alappat* inquiry simply requires an examination of the contested claims to see if the claimed subject matter as a whole is a disembodied mathematical concept representing nothing more than a "law of nature" or an "abstract idea," or if the mathematical concept has been reduced to some practical application rendering it "useful." Id. at 1544, 31 USPQ2d at 1557.

In *Alappat*, we held that more than an abstract idea was claimed because the claimed invention as a whole was directed toward forming a specific machine that produced the useful, concrete, and tangible result of a smooth waveform display. See id. at 1544, 31 USPQ2d at 1557.

In both *Alappat* and *State Street*, the claim was for a machine that achieved certain results. In the case before us, because *Excel* does not own or operate the facilities over which its calls are placed, *AT&T* did not charge *Excel* with infringement of its apparatus claims, but limited its infringement charge to the specified method or process claims. Whether stated implicitly or explicitly, we consider the scope of Section 101 to be the same regardless of the form - machine or process - in which a particular claim is drafted. See, *1358 e.g., *In re Alappat*, 33 F.3d at 1581, 31 USPQ2d at 1589 (Rader, J., concurring) ("Judge Rich, with whom I fully concur, reads *Alappat's* application as claiming a machine. In fact, whether the invention is a process or a machine is irrelevant. The language of the Patent Act itself, as well as Supreme Court rulings, clarifies that *Alappat's* invention fits comfortably within 35 U.S.C. 101 whether viewed as a process or a machine."); *State Street*, 149 F.3d at 1372, 47 USPQ2d at 1600 ("[F]or the purposes of a Section 101 analysis, it is of little relevance whether claim 1 is directed to a 'machine' or a 'process,'...."). Furthermore, the Supreme Court's decisions in *Diehr*, *Benson*, and *Flook*, all of which involved method (i.e., process) claims, have provided and supported the principles which we apply to both machine-- and process-type claims. Thus, we are comfortable in applying our reasoning in *Alappat* and *State Street* to the method claims at issue in this case.

C.

In light of this review of the current understanding of the "mathematical algorithm" exception, we turn now to the arguments of the parties in support of and in opposition to the trial court's judgment. We note that, at the time the trial court made its decision, that court did not have the benefit of this court's explication in *State Street* of the mathematical algorithm issue.

As previously explained, *AT&T's* claimed process employs subscribers' and call recipients' PICs as data, applies Boolean algebra to those data to determine the value of the PIC indicator, and applies that value through switching and recording mechanisms to create a signal useful for billing purposes. In *State Street*, we held that the processing system there was patentable subject matter because the system takes data representing discrete dollar amounts through a series of mathematical calculations to determine a final share price - a useful, concrete, and tangible result. See 149 F.3d at 1373, 47 USPQ2d at 1601.

In this case, *Excel* argues, correctly, that the PIC indicator value is derived using a simple mathematical principle (p and q). But that is not determinative because *AT&T* does not claim the Boolean principle as such or attempt to forestall its use in any other application. It is clear from the written description of the '184 patent that *AT&T* is only claiming a process that uses the Boolean principle in order to determine the value of the PIC indicator. The PIC indicator represents information about the call recipient's PIC, a useful, non-abstract result that facilitates differential billing of long- distance calls made by an IXC's subscriber. Because the claimed process applies the Boolean principle to produce a useful, concrete, tangible result without pre-empting other uses of the mathematical principle, on its face the claimed process comfortably falls within the scope of Section 101. See *Arrhythmia Research Technology, Inc. v. Corazonix Corp.*, 958 F.2d 1053, 1060, 22 USPQ2d 1033, 1039 (Fed.Cir.1992) ("That the product is numerical is not a criterion of whether the claim is directed to statutory subject matter.").

Excel argues that method claims containing mathematical algorithms are patentable subject matter only if there is a "physical transformation" or conversion of subject matter from one state into another. The physical transformation language appears in *Diehr*, see 450 U.S. at 184, 101 S.Ct. 1048 ("That respondents' claims involve the transformation of an article, in this case raw, uncured synthetic rubber, into a different state or thing cannot be disputed."), and has been echoed by this court in *Schrader*, 22 F.3d at 294, 30 USPQ2d at 1458 ("Therefore, we do not find in the claim any kind of data transformation.").

The notion of "physical transformation" can be misunderstood. In the first place, it is not an invariable requirement, but merely one example of how a mathematical algorithm may bring about a useful application. As the Supreme Court itself noted, "*1359 when [a claimed invention] is performing a function which the patent laws were designed to protect (e.g., transforming or reducing an article to a different state or thing), then the claim satisfies the requirements of Section 101." *Diehr*, 450 U.S. at 192, 101 S.Ct. 1048 (emphasis added). The "e.g." signal denotes an example, not an exclusive requirement.

This understanding of transformation is consistent with our earlier decision in *Arrhythmia*, 958 F.2d 1053, 22 USPQ2d 1033 (Fed.Cir.1992). *Arrhythmia's* process claims included various mathematical formulae to analyze electrocardiograph signals to determine a specified heart activity. See id. at 1059, 22 USPQ2d at 1037-38. The *Arrhythmia* court reasoned that the method claims qualified as statutory subject matter by noting that the steps transformed physical, electrical signals from one form into another form - a number representing a signal related to the patient's heart activity, a non-abstract output. See id., 958 F.2d at 1059, 22 USPQ2d at 1038. The finding that the claimed process "transformed" data from one "form" to another simply confirmed that *Arrhythmia's* method claims satisfied Section 101 because the mathematical algorithm included within the process was applied to produce a number which had specific meaning - a useful, concrete, tangible result - not a mathematical abstraction. See id. at 1060, 22 USPQ2d at 1039.

Excel also contends that because the process claims at issue lack physical limitations set forth in the patent, the claims are not patentable subject matter. This argument reflects a misunderstanding of our case law. The cases cited by *Excel* for this proposition involved machine claims written in means-plus-function language. See, e.g., *State Street*, 149 F.3d at 1371, 47 USPQ2d at 1599; *Alappat*, 33 F.3d at 1541, 31 USPQ2d at 1554-55. Apparatus claims written in this manner require supporting structure in the written description that corresponds to the claimed "means" elements. See 35 U.S.C. s 112, para. 6 (1994). Since the claims at issue in this case are directed to a process in the first instance, a structural inquiry is unnecessary.

The argument that physical limitations are necessary may also stem from the second part of the Freeman-Walter-Abele test, [FN4] an earlier test which has been used to identify claims thought to involve unpatentable mathematical algorithms. That second part was said to inquire "whether the claim is directed to a mathematical algorithm that is not applied to or limited by physical elements." *Arrhythmia*, 958 F.2d at 1058, 22 USPQ2d at 1037. Although our en banc *Alappat* decision called this test "not an improper analysis," we then pointed out that "the ultimate issue always has been whether the claim as a whole is drawn to statutory subject matter." 33 F.3d at 1543 n. 21, 31 USPQ2d at 1557 n. 21. Furthermore, our recent *State Street* decision questioned the continuing viability of the Freeman-Walter- Abele test, noting that, "[a]fter *Diehr* and *Chakrabarty*, the Freeman-Walter--Abele test has little, if any, applicability to determining the presence of statutory subject matter." 149 F.3d at 1374, 47 USPQ2d at 1601. Whatever may be left of the earlier test, if anything, this type of physical limitations analysis seems of little value because "after *Diehr* and *Alappat*, the mere fact that a claimed invention involves inputting numbers, calculating numbers, outputting numbers, and storing numbers, in and of itself, would not render it nonstatutory subject matter, unless, of course, its operation does not produce a 'useful, concrete and tangible result.' " Id. at 1374, 47 USPQ2d at 1602 (quoting *Alappat*, 33 F.3d at 1544, 31 USPQ2d at 1557).

> FN4. See *In re Freeman*, 573 F.2d 1237, 197 USPQ 464 (CCPA 1978), as modified by *In re Walter*, 618 F.2d 758, 205 USPQ 397(CCPA 1980), and In re Abele, 648 F.2d 902, 214 USPQ 682 (CCPA 1982).

Because we focus on the inquiry deemed "the ultimate issue" by *Alappat*, rather than on the physical limitations inquiry of *1360 the Freeman-Walter-Abele test, we find the cases cited by *Excel* in support of its position to be inapposite. For example, in *In re Grams*, the court applied the Freeman-Walter-Abele test and concluded that the only physical step in the claimed process involved data-gathering for the algorithm; thus, the claims were held to be directed to unpatentable subject matter. See 888 F.2d 835, 839, 12 USPQ2d 1824, 1829 (Fed.Cir.1989). In contrast, our inquiry here focuses on whether the mathematical algorithm is applied in a practical manner to produce a useful result. *In re Grams* is unhelpful because the panel in that case did not ascertain if the end result of the claimed process was useful, concrete, and tangible.

Similarly, the court in *In re Schrader* relied upon the Freeman-Walter-Abele test for its analysis of the method claim involved. The court found neither a physical transformation nor any physical step in the claimed process aside from the entering of data into a record. See 22 F.3d at 294, 30 USPQ2d at 1458. The *Schrader* court likened the data-recording step to that of data-gathering and held that the claim was properly rejected as failing to define patentable subject matter. See id. at 294, 296, 30 USPQ2d at 1458-59. The focus of the court in *Schrader* was not on whether the mathematical algorithm was applied in a practical manner since it ended its inquiry before looking to see if a useful, concrete, tangible result ensued. Thus, in light of our recent understanding of the issue, the *Schrader* court's analysis is as unhelpful as that of *In re Grams*.

Finally, the decision in *In re Warmerdam*, 33 F.3d 1354, 31 USPQ2d 1754 (Fed.Cir.1994) is not to the contrary. There the court recognized the difficulty in knowing exactly what a mathematical algorithm is, "which makes rather dicey the determination of whether the claim as a whole is no more than that." Id. at 1359, 31 USPQ2d at 1758. Warmerdam's claims 1-4 encompassed a method for controlling the motion of objects and machines to avoid collision with other moving or fixed objects by generating bubble hierarchies through the use of a particular mathematical procedure. See id. at 1356, 31 USPQ2d at 1755-56. The court found that the claimed process did nothing more than manipulate basic mathematical constructs and concluded that "taking several abstract ideas and manipulating them together adds nothing to the basic equation"; hence, the court held that the claims were properly rejected under Section 101. Id. at 1360, 31 USPQ2d at 1759. Whether one agrees with the court's conclusion on the facts, the holding of the case is a straightforward application of the basic principle that mere laws of nature, natural phenomena, and abstract ideas are not within the categories of inventions or discoveries that may be patented under Section 101.

D.

In his dissent in *Diehr*, Justice Stevens noted two concerns regarding the Section 101 issue, and to which, in his view, federal judges have a duty to respond:

> First, the cases considering the patentability of program-related inventions do not establish rules that enable a conscientious patent lawyer to determine with a fair degree of accuracy which, if any, program-related inventions will be patentable. Second, the inclusion of the ambiguous concept of an "algorithm" within the "law of nature" category of unpatentable subject matter has given rise to the concern that almost any process might be so described and therefore held unpatentable.

Diehr, 450 U.S. at 219, 101 S.Ct. 1048 (Stevens, J., dissenting).

Despite the almost twenty years since Justice Stevens wrote, these concerns remain important. His solution was to declare all computer-based programming unpatentable. That has not been the course the law has taken. Rather, it is now clear that computer-based programming constitutes patentable subject matter so long as the basic requirements of Section 101 are met. Justice Stevens's concerns can be addressed within that framework.

*1361 His first concern, that the rules are not sufficiently clear to enable reasonable prediction of outcomes, should be less of a concern today in light of the refocusing of the Section 101 issue that *Alappat* and *State Street* have provided. His second concern, that the ambiguous concept of "algorithm" could be used to make any process unpatentable, can be laid to rest once the focus is understood to be not on whether there is a mathematical algorithm at work, but on whether the algorithm-containing invention, as a whole, produces a tangible, useful, result.

In light of the above, and consistent with the clearer understanding that our more recent cases have provided, we conclude that the district court did not apply the proper analysis to the method claims at issue. Furthermore, had the court applied the proper analysis to the stated claims, the court would have concluded that all the claims asserted fall comfortably within the broad scope of patentable subject matter under Section 101. Accordingly, we hold as a matter of law that *Excel* was not entitled to the grant of summary judgment of invalidity of the '184 patent under Section 101.

Since the case must be returned to the trial court for further proceedings, and to avoid any possible misunderstandings as to the scope of our decision, we note that the ultimate validity of these claims depends upon their satisfying the other requirements for patentability such as those set forth in 35 U.S.C. ss 102, 103, and 112. Thus, on remand, those questions, as well as any others the parties may properly raise, remain for disposition.

CONCLUSION

The district court's summary judgment of invalidity is reversed, and the case is remanded for further proceedings consistent with this opinion.

REVERSED and REMANDED.

Appendix 3

e-Patents: Seven Examples for Corporate Financing

Seven examples of recent e-patents that we obtained for clients are discussed below. They illustrate different types of e-patents that are being obtained today. (E-patents are often voluminous, so we are reproducing here only the first page of each patent, with the abstract and identifying information. The complete patents are public documents, and can be obtained off the Web and elsewhere.) Of course, other types of e-patents are also obtainable.

I am happy to point out that applications for each of these patents were filed at the Patent Office before the Federal Circuit handed down the path breaking *State Street Bank* case (discussed in Chapter 2 of this book), which aggressively supported this patent strategy. *State Street Bank & Trust Co. v. Signature Financial Group, Inc.*, 149 F.3rd 1368, 47 U.S.P.Q.2d 1596 (Fed. Cir. 1998). I expected that the case would be decided as it was, and in the mean time, I sought early priority dates for the clients, without waiting for the decision.

These examples include patents for:

(1) internet services (U.S. Patent No. 5,963,951),

(2) a graphical user interface (GUI), that may be used on the Web or a LAN (Local Area Network) or a VPN (Virtual Private Network) (U.S. Patent No. 5,956,024),

(3) an internet gateway for remote network management (U.S. Patent No. 5,742,762) (the gateway may be embodied as a strictly software product, or as a hardware product with embedded software),

(4) a distributed computer data processing system with graphical user interface (GUI) that is embodied as a large software package (U.S. Patent 5,696,906),

(5) a second patent related to '906 that obtains 50 additional claims for the same invention (U.S. Patent No. 5,884,284),

(6) telecom satellite control software (U.S. Patent No. 5,978,363), and

(7) data analysis and display software (U.S. Patent No. 5,438,985).

Reviewing these sample patents leads us to several observations that illustrate several trends in e-patents discussed in this book.

The first observation is that e-patents were being obtained by the end of the 90's that were thought by many to be unobtainable and unenforceable at the beginning of the 90's. However, these e-patents are now clearly within the mainstream of practice and protected by the Federal Circuit case law and Patent Office procedures and practices that are expressed in Patent Office Guidelines in the Manual of Patent Examining Procedure. Further, as it is discussed in Chapters 1, 2, and

3 of this book, these patents are being enforced in the courts against infringers.

Perhaps a more striking observation comes from looking at the chain of owners of these example e-patents. All but one of the original owners were smaller or privately held companies that developed cutting edge e-patent strategies, and were shortly thereafter acquired by major players in their industries and/or went public. Continental Cablevision, Inc. was a privately held company that is now the core of MediaOne, Inc. (listed on the New York Stock Exchange). Telogy Networks, Inc. was a private start-up that is now a wholly owned subsidiary of Texas Instruments (listed on the New York Stock Exchange). Synectics Medical, Inc. was a privately owned Swedish company, which later did an IPO in Sweden, and is now a subsidiary of Medtronic, Inc. (listed on the New York Stock Exchange). This is indicative of the trend established in the 90's that sophisticated investors have become interested in aggressive e-patent strategies for their target companies. E-patent strategies are being recognized by the financial community as adding huge value to the owners, and are assuming make-or-break status for more technology deals.

In a sense, investments in the newer companies in these industries that have yet to develop large profit streams are investments in "investment grade" patent portfolios, where they can be found. Indeed, in the intangible world of the Internet and e-commerce services and software, younger high growth companies with only perspective profits, have little in the way of "property" aside from any intellectual property. Clearly, by the turn of the millennium, e-patents have established a critical role to facilitate financing, and are often a necessary condition for successful financing.

It is also interesting to note that many e-patents are developed by small or private companies that are later acquired by large public players in these industries. Small players sometimes are more able to innovate internally, than to acquire innovation from other corporate players. Larger players sometimes are better able to acquire innovative technology by acquisition than by internal development. However, this is changing in these technologies, and larger players are now increasingly protecting internally developed intellectual property, as well as acquiring intellectual property from others.

(1)
U.S. Patent No. 5,963,951
for an Internet Service,
to Movo Media, Inc.

United States Patent [19]
Collins

[11] Patent Number: 5,963,951
[45] Date of Patent: Oct. 5, 1999

[54] COMPUTERIZED ON-LINE DATING SERVICE FOR SEARCHING AND MATCHING PEOPLE

[75] Inventor: Gregg Collins, Los Angeles, Calif.

[73] Assignee: Movo Media, Inc., Los Angeles, Calif.

[21] Appl. No.: 08/885,199

[22] Filed: Jun. 30, 1997

[51] Int. Cl.[6] G06F 17/30
[52] U.S. Cl. 707/102; 707/3
[58] Field of Search 455/2, 4.2; 707/1, 707/2, 3, 4, 5, 100, 102; 364/282.1, 974; 379/201

[56] **References Cited**

U.S. PATENT DOCUMENTS

Re. 30,579	4/1981	Goldman et al. .
Re. 30,580	4/1981	Goldman et al. .
4,307,266	12/1981	Messina .
4,348,744	9/1982	White 395/200.83
4,427,848	1/1984	Tsakanikas .
4,649,563	3/1987	Riskin .
4,650,927	3/1987	James .
4,677,659	6/1987	Dargan .
4,737,980	4/1988	Curtin et al. .
4,817,129	3/1989	Riskin .
4,845,739	7/1989	Katz .
4,866,759	9/1989	Riskin .
4,918,721	4/1990	Hashimoto .
4,930,150	5/1990	Katz .
4,932,046	6/1990	Katz et al. .
4,939,773	7/1990	Katz .
4,975,945	12/1990	Carbullido .
4,987,590	1/1991	Katz .
5,014,298	5/1991	Katz .
5,016,270	5/1991	Katz .
5,031,206	7/1991	Riskin .
5,048,075	9/1991	Katz .
5,091,933	2/1992	Katz .
5,109,404	4/1992	Katz et al. .
5,218,631	6/1993	Katz .
5,224,153	6/1993	Katz .
5,255,309	10/1993	Katz .
5,259,023	11/1993	Katz .
5,297,197	3/1994	Katz .
5,337,347	8/1994	Halstead-Nussloch et al. .
5,339,358	8/1994	Danish et al. .
5,347,306	9/1994	Nitta 345/332
5,351,285	9/1994	Katz .
5,359,645	10/1994	Katz .
5,365,575	11/1994	Katz .
5,392,338	2/1995	Danish et al. .
5,442,688	8/1995	Katz .
5,495,284	2/1996	Katz .
5,548,634	8/1996	Gahang et al. .
5,553,120	9/1996	Katz .
5,561,707	10/1996	Katz .
5,764,736	6/1998	Shachar et al. 379/93.09
5,775,695	7/1998	Byers 273/161
5,796,395	8/1998	De Hond 345/331
5,802,156	9/1998	Felger 379/112
5,894,556	4/1999	Grimm et al. 463/42

OTHER PUBLICATIONS

Match.Com, Version 3.0, Electric Classified, PR Newswire San Francisco, Oct. 30, 1995.

Primary Examiner—Paul R. Lintz
Assistant Examiner—Ella Colbert
Attorney, Agent, or Firm—Stephen C. Glazier

[57] **ABSTRACT**

In an on-line dating service, a database of subscriber information is searched to find at least one subscriber matching user search criteria. The subscriber information includes preferences of subscribers to the service. Personal preferences for a user are obtained as search criteria. The personal preferences and the subscriber information include at least: a gender preference; a geographic location preference; an age preference; appearance preferences; religious belief preferences; educational level preferences; and a goal preference, and the goal preference is one of "romance"; "friendship" or "a walk on the wild side", the geographic location preference is at least one of a postal code, a country, a city, a suburb, a block, or a street. The subscriber information includes a date of last payment of a use fee by each subscriber and a date of last updating of a personal profile by each subscriber. The database is repeatedly searched for records matching the personal preferences of the user by at least a percentage match parameter value. The percentage match parameter is reduced by a value of say 10%, until at least a required number of matching records are found.

49 Claims, 7 Drawing Sheets

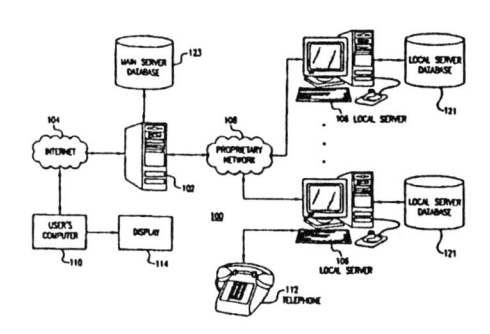

e-Patents: Seven Examples for Financing

(2)
U.S. Patent No. 5,956,024
for a Graphical User Interface (GUI),
to Continental Cablevision, Inc. (now MediaOne, Inc.)

US005956024A

United States Patent [19]
Strickland et al.

[11] Patent Number: 5,956,024
[45] Date of Patent: *Sep. 21, 1999

[54] **GRAPHICAL USER INTERFACE FOR CUSTOMER SERVICE REPRESENTATIVES FOR SUBSCRIBER MANAGEMENT SYSTEMS**

[75] Inventors: Marshall Strickland; Rob Strickland, both of Boston, Mass.

[73] Assignee: Continental Cablevision, Inc., Boston, Mass.

[*] Notice: This patent issued on a continued prosecution application filed under 37 CFR 1.53(d), and is subject to the twenty year patent term provisions of 35 U.S.C. 154(a)(2).

[21] Appl. No.: 08/659,323

[22] Filed: Jun. 6, 1996

Related U.S. Application Data

[60] Provisional application No. 60/002,045, Aug. 8, 1995.
[51] Int. Cl.⁶ G06F 9/00
[52] U.S. Cl. 345/327; 348/8
[58] Field of Search 345/326, 327, 345/329; 348/8, 10, 12, 13

[56] **References Cited**

U.S. PATENT DOCUMENTS

Re. 32,632	3/1988	Atkinson	345/146 X
4,698,624	10/1987	Barker	345/157
4,772,882	9/1988	Mical	345/146
5,115,504	5/1992	Belove	707/100
5,185,857	2/1993	Rozmanith	707/509
5,204,947	4/1993	Bernstein	345/357
5,220,675	6/1993	Padawer	345/333
5,262,761	11/1993	Scandura	345/346 X
5,297,249	3/1994	Bernstein	345/356
5,307,086	4/1994	Griffin	345/146
5,309,509	5/1994	Cocklin	379/165
5,317,687	5/1994	Torres	345/349
5,335,320	8/1994	Iwata	395/704
5,347,627	9/1994	Hoffmann	345/334
5,347,629	9/1994	Barrett	345/334
5,349,658	9/1994	O'Rourke	345/349
5,371,844	12/1994	Andrew	345/334
5,374,924	12/1994	McKiel, Jr.	345/340 X
5,384,910	1/1995	Torres	345/352
5,388,202	2/1995	Squires	345/334
5,404,441	4/1995	Satoyama	345/334
5,414,806	5/1995	Richards	345/349 X
5,414,836	5/1995	Baer	395/183.14
5,416,508	5/1995	Sakuma et al.	348/10
5,416,890	5/1995	Beretta	345/440 X
5,424,140	6/1995	Bloomfield .	
5,430,836	7/1995	Wolf	345/335
5,436,637	7/1995	Gayraud	345/346
5,524,195	6/1996	Clanton, III et al.	395/327
5,539,822	7/1996	Lett	348/13
5,592,212	1/1997	Handelman	348/8

Primary Examiner—Raymond J. Bayerl
Assistant Examiner—Cao H. Nguyen
Attorney, Agent, or Firm—Stephen C. Glazier

[57] **ABSTRACT**

The present invention is a graphical user interface ("GUI") for customer service representatives ("CSRs") for subscriber management systems ("SMS") for telecommunications service providers, including cable television multiple system operators ("MSOs"). A screen is provided with menus of icons for various functions at the customer service workstation ("CSW") for the CSRs. Sections of the screen provide presentation of various data to the CSR from the SMS.

12 Claims, 4 Drawing Sheets

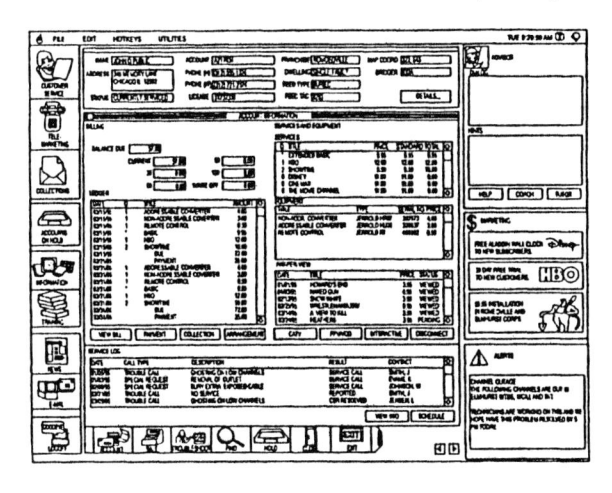

e-Patents: Seven Examples for Financing

155

(3)
U.S. Patent No. 5,742,762
for an Internet Gateway for Remote Network Management,
to Telogy Networks, Inc. (now a subsidiary of Texas Instruments, Inc.)

United States Patent [19]
Scholl et al.

[11] Patent Number: 5,742,762
[45] Date of Patent: Apr. 21, 1998

[54] NETWORK MANAGEMENT GATEWAY

[75] Inventors: Thomas H. Scholl; William E. Witowsky, both of Gaithersburg, Md.

[73] Assignee: Telogy Networks, Inc., Germantown, Md.

[21] Appl. No.: 444,483

[22] Filed: May 19, 1995

[51] Int. Cl.[6] G06F 13/00
[52] U.S. Cl. 395/200.3; 395/200.57
[58] Field of Search 395/200.01, 200.09, 395/200.11, 200.12, 370; 370/85.13

[56] References Cited

U.S. PATENT DOCUMENTS

5,327,544	7/1994	Lee et al.	395/500
5,491,693	2/1996	Britton et al.	370/85.13
5,491,796	2/1996	Wanderer et al.	395/200.09
5,508,732	4/1996	Bottmley et al.	348/7
5,530,852	6/1996	Meske, Jr. et al.	395/600
5,533,116	7/1996	Vestermen	379/243
5,559,800	9/1996	Mousseau et al.	370/85.13
5,581,558	12/1996	Horney, II et al.	370/401

Primary Examiner—Ayaz R. Sheikh
Attorney, Agent, or Firm—Stephen C. Glazier

[57] ABSTRACT

The present invention provides network management of a network or multiple networks, using a Web client, and including multimedia and hypermedia capability. The present invention provides a unified, remote, graphical, transparent interface for Web users, working at a Web client, to a variety of managed networks. The present invention receives requests from a Web client forwarded by a Web server and interacts with the managed networks and their associated objects to obtain information. The present invention then converts this information in real time to hypermedia document format in HTTP and HTML, and transmits this information to the Web client via the Web server, appearing to the client as information in a Web file. This permits a Web user to manage multiple networks and access multiple networks via a single Web client, thus providing a unification of the management interface for dissimilar managed networks, and devices.

18 Claims, 6 Drawing Sheets

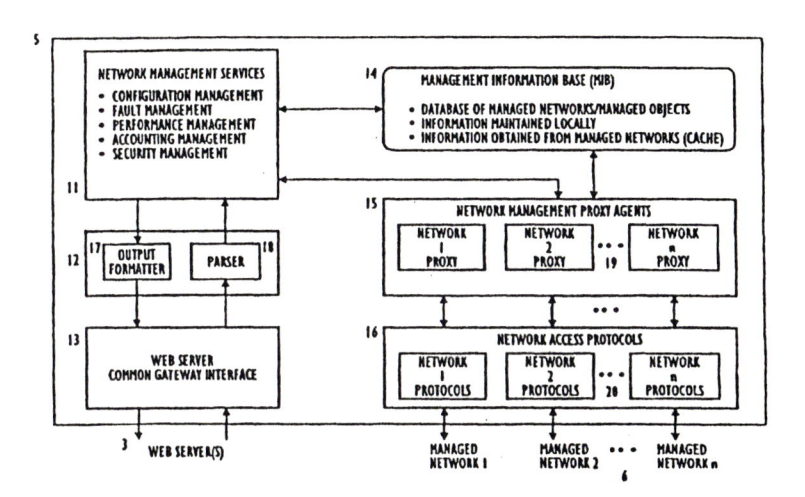

e-Patents: Seven Examples for Financing

159

(4)
U.S. Patent No. 5,696,906
for a Distributed Computer Data System with a GUI,
to Continental Cablevision, Inc. (now MediaOne, Inc.)

United States Patent [19]
Peters et al.

[11] Patent Number: 5,696,906
[45] Date of Patent: Dec. 9, 1997

[54] TELECOMMUNICAION USER ACCOUNT MANAGEMENT SYSTEM AND METHOD

[75] Inventors: J. Michael Peters; Barry Battista; Christopher Brown, all of Boston, Mass.

[73] Assignee: Continental Cablevision, Inc., Boston, Mass.

[21] Appl. No.: 401,602

[22] Filed: Mar. 9, 1995

[51] Int. Cl.^{6} G06F 17/60
[52] U.S. Cl. 395/234
[58] Field of Search 380/5; 379/121; 395/201, 207, 226, 230, 234, 240

[56] **References Cited**

U.S. PATENT DOCUMENTS

4,346,442	8/1982	Musmanno .
4,422,459	12/1983	Simson .
4,684,980	8/1987	Rast et al. .
4,754,426	6/1988	Rast et al. .
4,761,684	8/1988	Clark et al. .
5,019,900	5/1991	Clark et al. .
5,089,885	2/1992	Clark .
5,166,976	11/1992	Thompson et al. .
5,185,794	2/1993	Thompson et al. .
5,202,929	4/1993	Lemelson .
5,206,722	4/1993	Kwan .
5,220,501	6/1993	Lawlor et al. .
5,243,647	9/1993	Parikh et al. .
5,251,324	10/1993	McMullan, Jr. .
5,267,312	11/1993	Thompson et al. .
5,280,572	1/1994	Case et al. .
5,283,819	2/1994	Glick et al. .
5,291,477	3/1994	Liew .
5,291,554	3/1994	Morales 380/5
5,303,229	4/1994	Withers et al. .
5,311,325	5/1994	Edwards et al. .
5,319,455	6/1994	Hoarty et al. .
5,321,541	6/1994	Cohen .
5,347,632	9/1994	Filepp et al. .
5,359,642	10/1994	Castro 379/121

OTHER PUBLICATIONS

"Subscriber Management and Billing System Analysis", for Continental Cablevision, Inc., prepared by Caspen Consulting, Inc., Jul. 1993.

Primary Examiner—Gail O. Hayes
Assistant Examiner—Steven R. Yount
Attorney, Agent, or Firm—Stephen C. Glazier

[57] **ABSTRACT**

The present is an integrated computerized system and method of telecommunication user account management. The invention creates, maintains, processes and analyzes data regarding individual users for telecommunication services. Billing for individual users is generated. The user data is analyzed and reports for all or part of the user data are prepared and generated. Ancillary functions are enabled, including word processing, editing, e-mail, and other functions. The invention is applicable to subscriber telecommunication services, and pay-for-use services, and the user may be a subscriber or a non-subscriber. The invention is applicable to multi-channel telecommunication services, or single channel multi-service telecommunications, or single channel single service telecommunications. Such telecommunication services may include cable television, telephone, video, audio, on-line databases, television, radio, music video, video juke box, pay-for-view, video-on-demand, interactive TV, home-shopping, video conferences, telephone conferences, interfacing to imaging systems, automatic telephone call charge-backs ("900" numbers), and other telecommunication services which may not yet be invented at this time. The current preferred embodiment of the invention is for subscriber account management for cable television services.

2 Claims, 14 Drawing Sheets

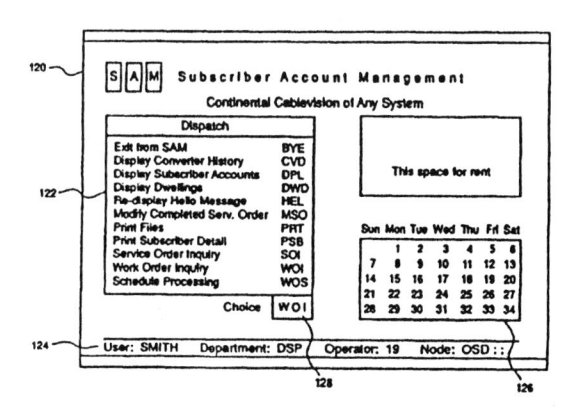

e-Patents: Seven Examples for Financing

(5)
U.S. Patent No. 5,884,284
for 50 additional claims for the invention of the preceding U.S. Patent No. 5,696,906 (for a Data System with a GUI),
to Continental Cablevision, Inc. (now MediaOne Group, Inc.)

US005884284A

United States Patent [19]
Peters et al.

[11] Patent Number: 5,884,284
[45] Date of Patent: Mar. 16, 1999

[54] TELECOMMUNICATION USER ACCOUNT MANAGEMENT SYSTEM AND METHOD

[75] Inventors: J. Michael Peters; Barry Battista; Christopher Brown, all of Boston, Mass.

[73] Assignee: Continental Cablevision, Inc., Boston, Mass.

[21] Appl. No.: 906,962

[22] Filed: Aug. 6, 1997

Related U.S. Application Data

[63] Continuation of Ser. No. 401,602, Mar. 9, 1995, Pat. No. 5,696,906.

[51] Int. Cl.⁶ G06F 17/60
[52] U.S. Cl. 705/30; 348/1; 348/3; 348/6; 348/7; 705/34; 705/400
[58] Field of Search 348/1, 3, 6, 7; 364/400; 705/1, 30, 34, 400

[56] **References Cited**

U.S. PATENT DOCUMENTS

4,346,442	8/1982	Musmanno 705/36
4,422,459	12/1983	Simson 600/515
4,684,980	8/1987	Rast et al. 348/7
4,695,880	9/1987	Johnson et al. 348/6
4,754,426	6/1988	Rast et al. 348/7
4,761,684	8/1988	Clark et al. 348/7
5,019,900	5/1991	Clark et al. 348/3
5,089,885	2/1992	Clark 348/7
5,166,976	11/1992	Thompson et al. 380/15
5,185,794	2/1993	Thompson et al. 380/17
5,202,929	4/1993	Lemelson 382/116
5,206,722	4/1993	Kwan 348/7
5,220,501	6/1993	Lawlor et al. 380/24
5,243,647	9/1993	Parika et al. 380/4
5,251,324	10/1993	McMullan, Jr. 455/2
5,267,312	11/1993	Thompson et al. 380/19
5,280,572	1/1994	Case et al. 369/49
5,283,819	2/1994	Glick et al. 379/93.01
5,287,270	2/1994	Hardy et al. 705/34
5,291,477	3/1994	Liew 370/238
5,291,554	3/1994	Morales 380/5
5,303,229	4/1994	Withers et al. 370/490
5,311,325	5/1994	Edwards et al. 348/5.5
5,319,455	6/1994	Hoarty et al. 348/7
5,321,541	6/1994	Cohen 359/127
5,325,290	6/1994	Cauffman et al. 705/34
5,347,632	9/1994	Filepp et al. 395/200.32
5,359,642	10/1994	Castro 379/121
5,592,551	1/1997	Lett et al. 380/20
5,696,906	12/1997	Peters et al. 380/5
5,727,055	3/1998	Ivie et al. 379/156

Primary Examiner—Edward R. Cosimano
Attorney, Agent, or Firm—Stephen C. Glazier

[57] **ABSTRACT**

The present invention is an integrated computerized system and method of telecommunication user account management. The invention creates, maintains, processes and analyzes data regarding individual users for telecommunication services. Billing for individual users is generated. The user data is analyzed and reports for all or part of the user data are prepared and generated. Ancillary functions are enabled, including word processing, editing, e-mail, and other functions. The invention is applicable to subscriber telecommunication services, and pay-for-use services, and the user may be a subscriber or a non-subscriber. The invention is applicable to multi-channel telecommunication services, or single channel multi-service telecommunications, or single channel single service telecommunications. Such telecommunication services may include cable television, telephone, video, audio, on-line databases, television, radio, music video, video juke box, pay-for-view, video-on-demand, interactive TV, home-shopping, video conferences, telephone conferences, interfacing to imaging systems, automatic telephone call charge-backs ("900" numbers), and other telecommunication services which may not yet be invented at this time. The current preferred embodiment of the invention is for subscriber account management for cable television services.

50 Claims, 14 Drawing Sheets

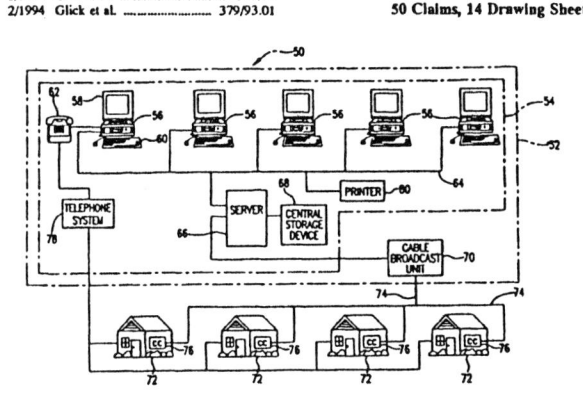

e-Patents: Seven Examples for Financing

167

(6)
U.S. Patent No. 5,978,363
for Telecom Satellite Control Software,
to Telogy Networks, Inc. (now a subsidiary of Texas Instruments, Inc.)

United States Patent [19]
Dimitrijevic et al.

[11] Patent Number: 5,978,363
[45] Date of Patent: *Nov. 2, 1999

[54] **SYSTEM AND METHOD FOR MULTI-DIMENSIONAL RESOURCE SCHEDULING**

[75] Inventors: **Dragomir D. Dimitrijevic; Dale Berlsford**, both of Germantown, Md.

[73] Assignee: **Telogy Networks, Inc.**, Germantown, Md.

[*] Notice: This patent issued on a continued prosecution application filed under 37 CFR 1.53(d), and is subject to the twenty year patent term provisions of 35 U.S.C. 154(a)(2).

[21] Appl. No.: **08/733,475**

[22] Filed: **Oct. 18, 1996**

[51] Int. Cl.⁶ H04B 7/185; H04B 7/212
[52] U.S. Cl. 370/319; 455/12; 455/13
[58] Field of Search 370/316, 317, 370/318, 319, 320, 325; 455/12, 13

[56] **References Cited**

U.S. PATENT DOCUMENTS

3,683,116	8/1972	Dill .
4,273,962	6/1981	Wolfe .
4,450,582	5/1984	Russell .
4,813,036	3/1989	Whitehead 370/325
4,870,642	9/1989	Nohara et al. 370/319
4,896,369	1/1990	Adams, Jr. .
4,995,096	2/1991	Isoe .
5,038,398	8/1991	Willis .
5,216,427	6/1993	Yan et al. 342/352
5,327,432	7/1994	Zein Al Abedeen .
5,363,374	11/1994	Zein Al Abedeen et al. 370/233
5,392,450	2/1995	Nossen .
5,430,732	7/1995	Lee et al. 370/319
5,448,621	9/1995	Knudsen .
5,485,464	1/1996	Strodtbeck et al. 370/319
5,519,404	5/1996	Cances et al. 370/319
5,526,404	6/1996	Wiedeman et al. 455/430
5,537,397	7/1996	Abramson .
5,537,406	7/1996	Binger .
5,572,530	11/1996	Chitre et al. 370/524
5,592,481	1/1997	Wiedeman et al. 370/316
5,625,624	4/1997	Rosen et al. 370/307
5,655,005	8/1997	Wiedeman et al. 370/320
5,669,062	9/1997	Olds et al. 455/509
5,708,965	1/1998	Courtney 455/13.4
5,749,044	5/1998	Natarajan et al. 455/13.1

Primary Examiner—Allen R. MacDonald
Assistant Examiner—James W. Mybre
Attorney, Agent, or Firm—Stephen C. Glazier; Pillsbury Madison & Sutro LLP

[57] **ABSTRACT**

A system and method is presented for multi-dimensional scheduling, unscheduling, and control of resources using a database approach. Indivisible resources, percentage divisible resources, and range divisible resources are scheduled, unscheduled, and controlled. Resources in telecommunication systems are scheduled, unscheduled, and controlled. Resources in satellite telecommunication systems are scheduled, unscheduled, and controlled.

5 Claims, 10 Drawing Sheets

Microfiche Appendix Included
(1 Microfiche, 23 Pages)

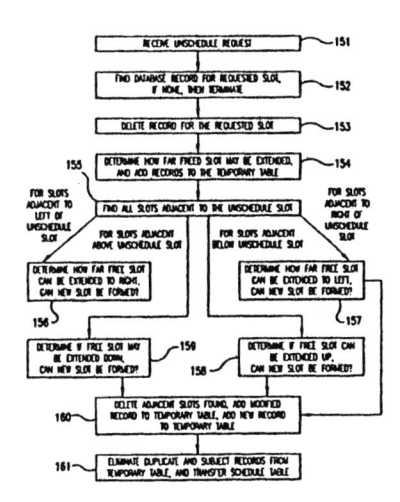

e-Patents: Seven Examples for Financing

171

(7)
U.S. Patent No. 5,438,985
for Data Analysis and Display Software,
to Synectics Medical, Inc. (now a subsidiary of Medtronic, Inc.)

United States Patent [19]
Essen-Moller

[11] Patent Number: 5,438,985
[45] Date of Patent: Aug. 8, 1995

[54] AMBULATORY RECORDING OF THE PRESENCE AND ACTIVITY OF SUBSTANCES IN GASTRO-INTESTINAL COMPARTMENTS

[75] Inventor: Anders Essen-Moller, Stockholm, Sweden

[73] Assignee: Synectics Medical, Incorporated, Irving, Tex.

[21] Appl. No.: 8,137

[22] Filed: Jan. 25, 1993

[51] Int. Cl.⁶ A61B 5/00
[52] U.S. Cl. 128/633; 128/635; 128/665
[58] Field of Search 128/632–635, 128/637, 664–666, 780

[56] **References Cited**

U.S. PATENT DOCUMENTS

2,162,656	6/1939	Warrington .
2,168,867	8/1939	George, 3rd .
2,857,915	10/1958	Sheridan .
3,373,735	3/1968	Gallagher .
3,480,003	11/1969	Crites .

(List continued on next page.)

FOREIGN PATENT DOCUMENTS

6673558	3/1983	European Pat. Off.	.... 128/634
0080680	6/1983	European Pat. Off.	.
0241644	10/1987	European Pat. Off.	.
0356603	11/1993	European Pat. Off.	.
2162656	6/1973	Germany	.
2453630	11/1980	Germany	.
3140265	4/1983	Germany	.
221635	5/1985	Germany	.
3523987	1/1987	Germany	.
4921789	2/1974	Japan	.... 128/635
7707275	1/1979	Netherlands	.
178028	11/1966	U.S.S.R.	.

OTHER PUBLICATIONS

Assorted promotional material by Synetics Medical, Inc.
Butcher et al., Digestion, 1992, vol. 53, pp. 142–148, "Use of an Ammonia Electrode for Rapid Quantification of *Helicobacter pylori* Urease: Its use in the Endoscopy Room and in the . . . ".
"Clinical relevance of ambulatory 24–hour . . . ", Vogten, et al., 1987, pp. 21-31 in Netherlands Journal of Medecine.
"Computerized Axial Manometry of the Esophagus", Bombeck, et al. in Annals of Surgery, vol. 206, No. 4, pp. 465–472, Oct. 1987.
"The laser motility sensor for long–term study of intra—esophageal pressure", Schneider et al., in Primary Motility Disorder of the Esophagus, Giuli et al., eds., pp. 64–69 1991.
The New Yorker, Sep. 20, 1993, T. Monmaney, "Marshall's Hunch".
"Oesophageal multipurpose monitoring probe", Baker et al., Anaesthesia, 1983, vol. 38, pp. 892–897.
World Wide Patent Monocrystant . . . (Brochure).
Digestive Diseases, Reprint, vol. 8, Suppl. 1, pp. 60–70, 1990, Scarpignato et al., "Simultaneous Measurement and Recording . . . ".
Hojgaard et al., "A New Method for Measurement of the Electrical Potential Difference Across the Stomach Wall", 1991. pp. 847–858.
"Ambulatory Monitoring of Gastric Emptying", Hoeft et al., May 16, 1993, American Assoc. of the Study of Live Diseases.

Primary Examiner—Lee S. Cohen
Assistant Examiner—Robert L. Nasser, Jr.
Attorney, Agent, or Firm—Stephen C. Glazier

[57] **ABSTRACT**

The present invention teaches a method and a system for ambulatory recording of the pH and the presence of various materials in compartments of the gastro-intestinal tract. The invention also reports the pH pattern in relation to the prevalence of the materials, and analyses to which degree such materials are in active or inactive states in their normal or foreign compartments. This is useful in situations, for example, when duodenal material is refluxed into the stomach and esophagus. The invention involves a gastro-intestinal catheter with a pH sensor and a combined light absorption and fluorescence sensor, a signal recorder and processor, and a written report producer.

4 Claims, 5 Drawing Sheets

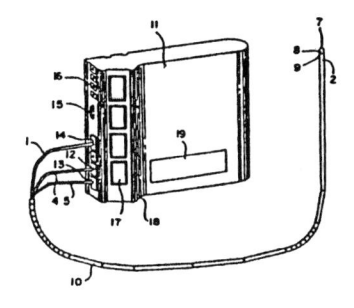

e-Patents: Seven Examples for Financing

175

Appendix 4

e-Patent Invention Disclosure Form

(Attach extra sheets if necessary.)
I.   TITLE OF THE INVENTION

II.   INVENTOR(S)

1. Name (first, middle initial, last):

Home Address:

Home Phone No.:

Work Phone No.:

Citizenship:

2. Name (first, middle initial, last):

Home Address:

Home Phone No.:

Work Phone No.:

Citizenship:

3. Name (first, middle initial, last):

Home Address:

Home Phone No.:

Work Phone No.:

Citizenship:

III.   DISCLOSURE EVENTS

Is the invention currently being used: Yes ___ No ___. If yes, where?

Has the invention been sold or offered for sale? Yes ___ No ___. If so, when and where?

Anticipated date of first disclosure to anyone under a Non-Disclosure Agreement.

Anticipated date of first public offer for sale.

Anticipated date of first public use or disclosure.

Anticipated date of first sale.

IV.   CONCEPTION OF THE INVENTION

Where and when was the invention first conceived?

Date of first drawing or written description?

Where can the first drawing or written description be found? Please attach a copy.

Does a prototype exist? Yes ___ No ___ If yes, where is it?

Did the prototype test out as functional: If not, why?

Is a prototype expected? Yes ___ No ___ If so, when?

Does any lab notes exist regarding the invention? If yes, please attach copies.

V. OTHER INFORMATION

Was the invention developed with third party assistance? If yes, who, when, what relationship?

Was the invention conceived, tested or constructed under a Government contract? If yes, please identify the contract, or attach a copy of the contract.

Identify **ALL** related publications, patents and patent applications of which you are aware:

Do you believe a prior art search would be warranted?
Yes ___ No ___
Note: If you believe that prior art of competitors exists, which prior art you cannot identify, and that prior art may suggest solving the problems that your inventions also solves, a prior art search is probably appropriate.

VI. DESCRIPTION OF THE INVENTION (USE ADDITIONAL SHEETS IF NEEDED)

Describe your invention, preferably with drawings of some type. Include enough detail to allow a person in the field to make the invention.

e-Patent Strategies

Identify the problems solved by the present invention.

Indicate the disadvantages of the prior methods or devices.

Describe the functional and/or structural differences between your solution and prior publicly known solutions.

Discuss the preferred environment in which your invention will work, and any modifications that would allow it to work in other environments.

What is the best way version to make or do the invention?

What variations in the invention could get substantially the same results, even if this other variation is in some ways inferior?

Identify the separate contributions of each inventor.

VII.   FOR INVENTIONS USING SOFTWARE

Please include flow charts, object diagrams, data structure diagrams, or other drawings showing the essential features of the invention.

e-Patent Invention Disclosure Form

Describe the software as a functional process with steps. What does the software do, and how does it do it.

What hardware and software platform does the software require to execute? What protocols are used, can be used, and which are preferred?

If the software system operates through a network, diagram the architecture of the network.

Describe any essential algorithms used by the software.

Does the software have an important or unique GUI (graphical user interface), or screen display, or icon? If so, provide a drawing.

Does complete source code exist for the software? If yes, where is it, or attach a copy. What language is the source code written in, and what other languages could be used? Which is preferred?

VIII. SIGNATURES

I hereby declare that the above statements are accurate and complete to the best of my knowledge and I assign here all rights to this invention to _____ [corporate employer].

A. Inventor's Signature

Print Name

Date

B. Inventor's Signature

Print Name

Date

C. Inventor's Signature

Print Name

Date

We, the individuals identified below, have read and understand the above referenced Invention Disclosure.

A. Witness Signature

Print Name

Date

e-Patent Invention Disclosure Form

B. Witness Signature

Print Name

Date

IX. **The following items are to be completed by the Engineering Manager or Supervisor of the above named inventor(s), whether or not such Manager or Supervisor is one of the inventors(s).**

In what existing product(s) will the invention be used? If no such product(s) yet exist, provide a brief description of potential product(s) incorporating this invention.

What is the potential market for products incorporating this invention?

Who are the potential competitors for such products?

What potential economic or strategic impact will this invention have on your company?

If the invention is software based, can the product be packaged as either software, or hardware with embedded software?

The undersigned Manager has read and understands the above Invention Disclosure

Manager's Signature

Date

Print Name

Work phone

e-Patent Invention Disclosure Form

Table of Authorities

1. CASES

In re Alappat
  33 F.3d 1526, 31 USPQ2d 1545 (Fed. Cir. 1992) 23, 30

Arrhythmia Research Technology, Inc. v. Corazonix Corp.
  958 F.2d 1053, 22 U.S.P.Q.2d 1033 (Fed. Cir. 1992) . 23, 30, 38

AT&T Corp. v. Excel Communications, Inc.
  172 F.3d 1352, 50 U.S.P.Q.2d 1447
  (Fed. Cir., April 14, 1999) 37, 39, 41

Diamond v. Diehr
  450 U.S. 175, 101 S.Ct. 1048, 67 L.Ed.2d 155 (1981) 31

Paine, Webber, Jackson & Curtis, Inc. v. Merrill, Lynch, Pierce, Fenner & Smith, Inc.
  564 F.Supp. 1358, 218 U.S.P.Q. 212 (D. Del. 1983) .. 23, 31, 46

In re Schrader
  22 F.3d 290, 30 U.S.P.Q.2d 1455 (Fed. Cir. 1944) 33

Stac Electronics v. Microsoft Corp.
  (D.C. C. Cal. CV-93-413-ER, 1994) 6, 8, 60, 61

State Street Bank & Trust Co. v. Signature Financial Group, Inc.
  149 F.3d 1368, 47 U.S.P.Q.2d 1596
  (Fed. Cir. 1998) 2, 23, 26, 28,
  31, 32, 33, 34
  37, 39, 41, 46

2. LEGISLATION

Economic Espionage Act of 1996
  110 Stat. 3488, October 11, 1996 .... 5, 37

3. U. S. PATENTS

U.S. Patent No. 5,193,056 .... 26, 28

4. NON-U.S. LAW

European Patent Convention
  Article 52 (2) (c) 40

Index

1952 Patent Act, 33

algorithm, decision making, 56; encryption, 56; mathematical, 38; program trading, 31
Amazon.com, 54, 55
artificial intelligence, 54
assignments, 11, 13-14; verification, 21
ATM (asynchronous transfer mode), 77

bandwidth, 78
banks, as patent holders, 4
BarnesandNoble.com, 54
biometrics, 98

Citibank, number of patents, 45
claims, for software, 28; mathematical algorithm, 38; means plus function, 29; method of doing business, 28, 34; patentable subject matter, examples, 27; process, 34; submarine, 68
confidentiality agreements, 66
continuations and continuations-in-part (CIPs), 56-57
Court of Patent Appeals, 33
culture of confidentiality. See trade secrets and confidentiality

Dell.com, 89
deregulation, 79
digital telecom devices, 53
Drugstore.com, 54
due diligence, 1-22; basics, 21-22; and industrial espionage, 4

e-commerce, 81; and consumers, 90; and regulation, 93; payment systems, 91-92; software, 91; trends 87-93
e-patents, definition, xi
e-tailing, See Internet, retailing applications.
European Patent Convention, 40
European Patent Office, Guidelines for Examination, 40
extranets, 54; trends, 81-85

Federal Economic Espionage Act, 5, 36
foreign patents, 21
Fromages.com, 55

GPS (global positioning system), 97
GUI (Graphical User Interface), 50-51; aesthetics, 52

home banking, number of patents, 45
hub and spoke patent, 32, 33, 36; and mutual funds, 26, 28

i-patents, definition, xi
IBM Corp., 39
infringement, 26, 60, 69; cease and desist letters, 21-22; potential, 15-16; risks, 36
intellectual property audits, 65-70; 71-76; and a business plan, 65; and industrial intelligence, 71-72; internal, 73-76
Internet, bookstore, 54; direct marketing, 92; distribution systems, 55; life insurance, 54; number of patents, 45-46; patent strategy, 88; protocols, 82-83; remote payment, 55-56; retailing applications, 55; secure access, 82; software profit models, 84; stock trading, 54; trends, 81-85
intranet, trends, 81-85
invent around strategy, 22, 56, 69, 72
invention-just-in-time, 22
invention-on-demand, 22
inventions, "new combination of old parts," 49-50; "Swiss

army knife," 51

Jobs, Steven, 52

killer app, 98

laches, 21-22
leap frog strategy, 72, 76
licenses, review of, 21
licensing, 10-11, 35, 75

Manual of Patent Examining Procedure, Section 706.03(a), 34
market share, 60, 61
Mastercard, number of patents, 44
medical devices, 17
Merrill Lynch, CMA patent, 35-36
Microsoft Corp., 6-9, 61, 63; antitrust litigation, 85

niche players, 79-80
novelty, 26

patentability, and tangible results, 30-31, 38; business method exception, 32-33; for financial services, 32; mathematical algorithm exception, 29-32; requirements, 32, 38-39; subject matter, 29, 34, 38
patent attorneys, 20, 67
patent drafting, 32
patent enforcement, 34
patenting, and banking services, 43-47; and program trading algorithm, 31; and the Internet, 43-47; for financial services, 24, 25, 28, 29, 36; for the insurance industry, 35; legally driven, 68; market driven, 67; method of doing business, 25-26; obsolency, 62; pro-software trends, 37, 39, 41; technology driven, 68

patent lineage charts, 75-76
patent litigation, 8, 46, 61, 63
patent portfolio, 13-14, 17-18, 19, 22; infringement, 26; ownership, 20
patent professionals. See patent attorneys
Patent Statute, Section 101, 32-33, 33-34, 37, 38, 39
patent strategies, for developed industries, 24; for financial services, 24, 25, 26; for telecom services, 24
personal digital assistants, 53, 85
picket fence strategy, 76
Priceline.com, 54, 55, 88
Primary Interchange Carrier (PIC), 38
prior art, 15, 18, 75
private switched telephone networks (PSTN), 77
product development, legally driven, 18; market driven, 18; technology driven, 18
proprietary patented manufacturer, 17

representations and warranties, 22

satellite wireless telecom, 99
Section 101 of the patent statute. See Patent Statute
service patent, xii
smart cards, number of patents, 46; trends, 95-96
software, and tangible results, 31; copyright protection, 8; for banking services, 43; functions and protocols, 53; new patent litigation, 8; ownership, 22; patentability, xii, 2, 24, 28, 67; patent portfolio, 34; patent protection, 8; piracy, 40-41; programs, 39-40; reverse engineering, 50
software patenting, 23-24, 23-36, 29, 50; 59; stages of denial, 59-63

Stac Electronics, 6-9, 61, 63

subject matter, categories, 32; practical utility, 32; statutory, 33, 37, 39; unpatentable, 30
submarine strategy, 57

technology, trends, 97-99
telecom services, 2, 50; trends, 77-80
title defects, 66
title opinion letter, 21
tollgate strategy, 57, 72
trade secrets and confidentiality, 21, 36, 66

Uniform Trade Secrets Act, 5

validity opinions, 20
video cams, 98
virtual genius, rules, 49-57

web browser, 84-85
web server, 84
wireless telecom, 98-99
Wright Brothers, 51

zero inventory, 89